Education Division 2020

AA009351

Held at the 2020 AIChE Annual Meeting

Online
16-20 November 2020

ISBN: 978-1-7138-2299-8

Printed from e-media with permission by:

Curran Associates, Inc.
57 Morehouse Lane
Red Hook, NY 12571

Printed with permission by Curran Associates, Inc. (2021)

For permission requests, please contact AIChE
at the address below.

AIChE
120 Wall Street, FL 23
New York, NY 10005-4020

Phone: (800) 242-4363
Fax: (203) 775-5177

www.aiche.org

Additional copies of this publication are available from:

Curran Associates, Inc.
57 Morehouse Lane
Red Hook, NY 12571 USA
Phone: 845-758-0400
Fax: 845-758-2633
Email: curran@proceedings.com
Web: www.proceedings.com

TABLE OF CONTENTS

617187 Developing Your Mentorship Program

Abstract:

Mentorship is a great way to involve and develop underclassmen in your chapter while fostering a tight-knit community within your ChE department. Come check out how the University of Michigan AIChE Chapter runs their mentorship program!

617191 LEAPing Online: Maintaining Technical Proficiency at Home

Abstract:

Like many other student chapters across the country, AIChE at UCLA's technical projects were put to a halt when administration announced that instruction during spring would be conducted remotely. Despite this, we continued supporting our members by running one project, the Lifetime Engineering Applications Project (LEAP), completely online. LEAP utilizes chemical engineering process simulators to model industrial chemical processes, while also providing students an ethical and economic perspective on industry. This presentation will explore our experience transitioning to an online technical project and how to adapt to changes in budget, participation, and scheduling. Starting with an explanation on the relevant experience LEAP provides to students pursuing careers in industry, we will express our motivation for moving online and what makes a project viable for online instruction. Afterwards, we will discuss maximizing the resources an online environment introduces, maintaining participation levels, and being as accommodating as possible for our project members.

617195 AIChE Mentoring Program: Your Experience Is Mine

Abstract:

The AIChE mentoring program is a project design to provide freshman year students with a personalized senior student mentor in order to motivate, guide and teach them practical studying methods that could guarantee a good academic performance as well as a personal growth within the community and to embrace AIChE values like leadership, social conscience and proactivity.

617194 Diversifying Technical Projects: The Chemical Engineering Introductory Project

Abstract:

The Chemical Engineering Introductory Project (ChIP) is a technical project that allows participants to learn about fundamental chemical engineering principles and apply them by building a small-scale continuous process system in a team-oriented environment. In past years, our members have built coffee machines, cooling towers, and soap batch reactors as their final project. ChIP is aimed toward incoming freshmen and transfer students, and since its introduction three years ago, AIChE at UCLA has seen an increase in student participation in technical projects and other areas of the club. Topics covered in the workshop will include strategies and difficulties faced in implementing the coffee machine project, connecting coffee machines to core engineering fundamentals, and overall advantages that ChIP has to offer.

617199 How to Run a Career Meeting: Interview Preparation, Resume Review, and LinkedIn Help

Abstract:

As a school that utilizes co-ops in their curriculum, Northeastern students have to be prepared for job interviews as early as the fall semester of their sophomore year. Because of this, the NU chapter of AIChE runs a meeting every semester that includes resume review, interview preparation, LinkedIn assistance, and professional headshots. Upperclassmen give the advice, some of whom have gone on more than twenty interviews for co-op jobs. We want to run a workshop at the National AIChE Conference to give student leaders resources, such as interview tips and resume advice that our E-board has compiled over the years to guide each chapter in hosting their own career advancement meeting. We will then explain how to efficiently run a meeting that will allow all attending students to receive the help they need for any of these critical professional development skills.

600369 Development of an Electroactive Platform for Detection of Virus Fusion to Host Membranes

Abstract:

Every year, new virus outbreaks are reported and have potential to cause pandemics with devastating socio-economic impacts, with the coronavirus disease 19 (COVID-19) serving as a prime example. Recent viral outbreaks in this decade have alluded to the importance of quick and accurate viral sensing strategies so that the infected individual can begin their treatment and limit contact with uninfected individuals quicker, mitigating the impact of an outbreak. PCR, one of the most common viral detection methods, is a specific and sensitive detection technique, but requires trained personnel and sophisticated machinery, which can slow response times. An ideal detection method should integrate the specificity and sensitivity with the speed necessary to diagnosis patients rapidly.

Enveloped viruses (such as influenza virus, coronavirus) enter host cells by binding to a receptor and then fusing its viral membrane to the host cell membrane to insert its genome. We propose a novel sensing platform that leverages the virus's propensity to bind specifically and fuse to a host cell membrane. We can couple a biomimetic membrane on top of an electrically conductive polymer (PEDOT:PSS) to probe the electrical responses of the system once viral fusion has occurred. Successful viral fusion to the membrane in a matter of minutes may cause a detectable change in electrical response, serving as the readout of a viral infection. Using influenza virus as a proof-of-principle, we first show that we can form the biomimetic membrane incorporating the influenza viral receptor on PEDOT:PSS. We then visualize influenza single virion fusion on PEDOT:PSS and demonstrate that the kinetics of this fusion event are comparable to influenza fusion on a non-conductive polymer. Lastly, using electrical impedance spectroscopy, we provide evidence that influenza viral fusion causes a detectable change in electrical response, demonstrating that it is possible to exploit viral membrane fusion as a diagnostic strategy.

599943 Engineering DNA-Based Materials for the Analysis of Live Single Cells

Abstract:

The ability to probe the composition of live cells is important for learning about chemical processes inside of cells as well as for creating diagnostic platforms based on molecular profiling. However, current strategies for detecting analytes in single living cells are severely limited both in terms of probe design and sensing strategy. While fluorescent proteins and nucleic acid-based aptamers have revolutionized live-cell imaging, these sensors must be genetically encoded into the cells and therefore have limited applications in a clinical setting. Many techniques require cell lysis or a bulk population of cells to yield a detectable quantity of the target analyte. Additionally, a major impediment to the development of intracellular probes is the inability of most probes to traverse the cell membrane. Consequently, these probes require the use of transfection reagents or physical methods that perturb the cell membrane, which in turn can interfere with the measurements by changing the analyte's expression profiles. Furthermore, existing strategies for signal transduction to indicate the presence of a given analyte, such as fluorophore-quencher-based systems and FRET-based methodologies, are limited and susceptible to false-positive signal. Therefore, the development of a versatile platform for intracellular analyte detection in live cells is an outstanding challenge.

To address this challenge, I have first developed DNA-based aptamers that utilize a novel sensing strategy. Specifically, these aptamers are chemically modified at a single base with a viscosity-sensitive dye. Target binding to the aptamer changes the conformation of the aptamer and forces the dye to intercalate (or FIT) between the DNA base pairs, turning its fluorescence on. This "FIT" strategy avoids false-positive signals common to most fluorophore-quencher-based systems, quantitatively outperforms FRET-based techniques, gives up to 20-fold fluorescence enhancement in buffer, and allows for the detection of analytes in complex biological milieu such as human serum.

To apply this technique to the analysis of intracellular analytes, I have then developed a new platform based on protein spherical nucleic acids (ProSNAs). ProSNAs are composed of a protein core surrounded by a DNA shell. The dense functionalization and radial arrangement of the DNA around the protein core allows efficient cellular uptake of the ProSNA without the use of transfection reagents or physical stimuli. By using a FIT-aptamer as the DNA sequence, this design allows for the detection of intracellular analytes. As an example, I demonstrate that using an i-motif sequence that undergoes pH-dependent conformational changes allows for the detection of intracellular pH. As more than 500 aptamers are known for various targets, our nano-construct is a powerful platform for the detection of intracellular analytes. Moreover, by using a functional protein as the core, it is possible to detect analytes for which aptamers do not exist and further expand the range of analytes that can be detected. Using glucose oxidase as the core, I show that glucose can be detected (for which is there is no known aptamer with a biologically relevant binding affinity) in 8 different cell lines. Taken together, these constructs represent a plug-and-play platform for detecting a wide variety of intracellular analytes with great potential in the field of biodiagnostics.

References

†Equal Contribution

Samanta, D.†; **Ebrahimi, S. B.**†; Mirkin, C. A., "*Nucleic-Acid Structures as Intracellular Probes for Live Cells,*" *Advanced Materials* **2020**, *32*, 1901743.

Ebrahimi, S. B.†; Samanta, D.†; Cheng, H. F.; Nathan, L. I.; Mirkin, C. A., "*Forced Intercalation (FIT)-Aptamers,*" *Journal of the American Chemical Society* **2019**, *141*, 13744.

Samanta, D.; Iscen, A.; Laramy, C. R.; **Ebrahimi, S. B.**; Bujold, K. E.; Schatz, G. C.; Mirkin, C. A., "*Multivalent Cation-Induced Actuation of DNA-Mediated Colloidal Superlattices,*" *Journal of the American Chemical Society* **2019**, *141*, 19973.

608845 Molecular Engineering of a Colorimetric Nanogel Sensor for Clinical Radiotherapy and Trauma Monitoring

Abstract:

Radiotherapy using ionizing radiation (e.g. X rays) remains a mainstay of treatment modalities for different types of cancers. Overexposure to radiation in the clinic or as part of an adverse event can induce tissue toxicity and damage, whereas underdosing can lead to poor efficacies of tumor ablation. Effective radiation sensors are critical in ensuring optimum delivery to patients undergoing radiotherapy and for patient stratification in cases of radiation exposure. Currently existing sensors possess several disadvantages in rapid radiation detection because of their fragility, sensitivity to light and heat, dose-rate dependence, long processing times, lack of conformation to tissue morphologies and / or high costs. Commonly used Nanodots^{TM} are expensive, require separate read-out devices and do not conform to tissue morphologies. We developed a novel colorimetric nanogel sensor, based on the formation of maroon-colored gold nanoparticles from their colorless precursor salt formulations, upon exposure to different levels and types of ionizing radiation. By measuring the intensity of color developed with a simple bench-top absorbance spectrophotometer we were able to detect doses typically administered in fractionated cancer radiotherapy (i.e. ranging from 0-5 Gy doses) and beyond. Topographical distribution of delivered radiation doses was qualitatively and quantitatively determined following addition of quenching agents, and the efficacy of the nanogel sensor was demonstrated for different sources of ionizing radiation including protons, electrons, photons and radioactive isotopes. The translational potential of the sensor was demonstrated using anthropomorphic phantoms and in live canine patients undergoing clinical radiotherapy treatments. Our results demonstrate a new generation of nanogel sensors for colorimetric detection of radiation for improving patient safety in clinical radiotherapy and trauma care.

609804 Unique Electrochemical Detection of Sepsis Using Triplex Biomarker Detection Panel with IL-6, IL-8 and IL-10 in Blood Plasma

Abstract:

Sepsis is a life-threatening disease that is caused by the body's extreme reaction to an infection which could lead to tissue damage, multiple organ failure or even death. It is one of the leading causes of death worldwide, estimated to affect more than 30 million people every year according to the world health organization. Despite scientific developments in health care technologies in the past three decades, a substantial challenge continues to persist for early rapid diagnosis of sepsis. Thorough and timely diagnosis plays a substantial role in increasing the survival chances for patients with adequate treatment, thus reducing mortality. However, current gold standard blood culture techniques require large patient sample volumes and provides delayed results (6 hours to up to 5 days). This may lead to clinical ambiguity in treatment measures, propelling the body's immune response to spiral uncontrollably into septic shock or severe sepsis. This fails to deliver the required speed and accuracy to enable accurate diagnosis. To overcome existing challenges, we tailored a specific biosensor to screen for sepsis as a rapid point-of-care (POC) technology enabling multiplexed detection capabilities. The novelty presented in this research combines the effectiveness of specific multiplexed biomarkers for detection of sepsis with electrochemical impedance spectroscopy (EIS) technique to enable clinically efficient decision-making process. This work leverages the detection of pro and anti-inflammatory cytokine biomarkers such as interleukin 6 (IL-6), interleukin-8 (IL-8) and interleukin-10 (IL-10) in blood plasma. Researchers suggests good correlation between timing of pro and inflammatory cytokine release to degree of sepsis, associated with deleterious effects. The developed label-free, point-of-care biosensor is surface engineered with specific capture probes that utilizes EIS to measure the capacitive impedance change reflecting binding interactions between the capture probe and target biomarker enabling simultaneous triplex detection. Specificity of the POC biosensor was validated using cross-reactive studies, which displayed insignificant interference from non-specific biomarkers. The biosensor also displays stable and repeatable performance. Key advantage of the triplex biosensor is the use of undiluted blood plasma volume of less than 40 μL, providing result within 15 minutes. Additionally, the triplex biosensor technology was validated with 20 patient samples that were clinically tested positive for sepsis. This unique sensor technology would allow medical practitioners to facilitate targeted interventions for septic patients as a rapid prognostic approach, preventing complications arriving from sepsis.

598873 Harnessing the Protein Corona Towards Carbon Nanotube-Based Sensor Design

Abstract:

Unpredictable protein adsorption on nanoparticles remains a considerable challenge towards effectively applying nanotechnologies, such as nanosensors, in biological environments. However, if well understood and controlled, proteins adsorbed on nanoparticles can be exploited to develop protein-based nanosensors.

Herein, we employ our knowledge of spontaneous "protein corona" formation on single-walled carbon nanotubes (SWCNTs) in biofluids towards rational nanosensor design based on protein-SWCNT complexation.[1] SWCNTs are implemented as the sensor platform due to their desirable optical and physical characteristics, including photostable fluorescence in the near-infrared tissue transparency window[2] and high aspect ratio that confers delivery capabilities.[3] We first study the protein corona composition formed on SWCNT-based probes within relevant biofluids such as blood plasma and cerebrospinal fluid.[4] Then, we take advantage of the intrinsic binding affinity of specific proteins on the carbon nanotube to selectively adsorb proteins that, in turn, bind to their natural ligand partners. Thus, the adsorbed protein acts as the recognition moiety that interacts with the environment and detects the analyte of interest, tethered to the SWCNT that serves as the signal transducer. This nanosensor platform combines the sensitive and selective binding capabilities of proteins for their ligand partners with the near-infrared fluorescence readout of SWCNTs that is modulated upon protein-ligand binding. In this manner, we demonstrate that although formation of the nanoparticle protein corona can impede the intended nanobiotechnology application, it also presents an opportunity to create improved protein-nanoparticle architectures by manipulating selective protein adsorption on the nanoparticle surface.

References

1. Pinals, R. L., Yang, D., Lui, A., Cao, W. & Landry, M. P. Corona Exchange Dynamics on Carbon Nanotubes by Multiplexed Fluorescence Monitoring. *J. Am. Chem. Soc.* (2019) doi:10.1021/jacs.9b09617.
2. Boghossian, A. A. *et al.* Near-Infrared Fluorescent Sensors based on Single-Walled Carbon Nanotubes for Life Sciences Applications. *ChemSusChem* **4**, 848–863 (2011).
3. Demirer, G. S. *et al.* High aspect ratio nanomaterials enable delivery of functional genetic material without DNA integration in mature plants. *Nat. Nanotechnol.* **14**, 456–464 (2019).
4. Pinals, R. L. *et al.* Protein Corona Composition and Dynamics on Carbon Nanotubes in Blood Plasma and Cerebrospinal Fluid. *bioRxiv* 2020.01.13.905356 (2020) doi:10.1101/2020.01.13.905356.

607484 Point-of-Care Cancer Biomarker Detection System Integrating Surface Acoustic Wave Streaming and Metal-Enhanced Fluorescence

Abstract:

Point-of-care cancer biomarker detection system integrating surface acoustic wave streaming and metal-enhanced fluorescence

Yuqi Huang, Shuangming Li and Venkat R. Bhethanabotla

yuqi@usf.edu

shuangming@mail.usf.edu

bhethana@usf.edu

Department of Chemical and Biomedical Engineering, University of South Florida, Tampa, Florida 33620-5350, USA

Abstract:

Immunofluorescence is one of the primary techniques for microbiological assays. It can help to visualize, identify, and even quantify the interaction between antibody and antigen, some even used for DNA and protein detection.[1,2] However, there is a clinical need of increasing its limit of detection, specificity and portability. In this work, a portable optical fluorescence system was designed to be combined with a direct digital synthesizer for surface acoustic wave (SAW) generation and metal-enhanced fluorescence (MEF) to lower the detection limit in immunofluorescence assays.[3,4] MEF was achieved by depositing silver nanoparticles with covered silica film, which can plasmonically enhance fluorescence intensity. Rayleigh wave streaming was generated on a piezoelectric device for surface acoustic wave streaming purpose, which can prevent nonspecific binding on the biological interface.[5] The miniaturized prototype, integrating acoustic wave removal of interfering proteins combined with optical detection, was developed to realize immune detection on a portable platform with limit of detection at nanogram per milliliter level in detecting carcinoembryonic antigen (CEA) from human plasma.

Recent work shows that the MEF substrate fabricated with the above method can lower the detection limit of CEA from PBS to 100pg/mL level. However, testing the detection limit for CEA from human plasma has raised to 500ng/mL. Non-specific proteins, dusts and electrolytes in human plasma are primary suspects to raise background noise which increases the limit of detection. Further tests show that by adding SAW streaming to the sample tested on MEF substrate, the detection limit brings back down to 50pg/mL. A reasonable inference to this detection limit drop is the removal of non-specific protein by streaming surface acoustic wave to the sample surface, where the non-specific absorbents are removed by shear force from the wave, and the background noise is reduced to lower the limit of detection for CEA from human plasma.

1. K. Aslan, I. Gryczynski, J. Malicka, E. Matveeva, J. R. Lakowicz, and C. D. Geddes, "Metal-enhanced fluorescence: an emerging tool in biotechnology," Current Opinion in Biotechnology, vol. 16, pp. 55-62, **2005**.
2. R. A. Goldsby, T. J. Kindt, B. A. Osborne, and J. Kuby, "Immunology," ed: New York: WH Freeman, **2003**.
3. S. Li and V. Bhethanabotla, "Design of a Portable Orthogonal Surface Acoustic Wave Sensor System for Simultaneous Sensing, Removal of Nonspecifically bound Proteins and Mixing", Sensors, vol. 19(18), **2019**.
4. Liu, J.; Li, S.; Bhethanabotla, V. R., Integrating Metal-Enhanced Fluorescence and Surface Acoustic Waves for Sensitive and Rapid Quantification of Cancer Biomarkers from Real Matrices. *ACS Sensors*, *3* (1), 222-229, **2018**.
5. S. Cular, D. W. Branch, V. R. Bhethanabotla, G. D. Meyer, and H. G. Craighead, "Removal of nonspecifically bound proteins on microarrays using surface acoustic waves," IEEE Sensors journal, vol. 8, pp. 314-320, **2008**.

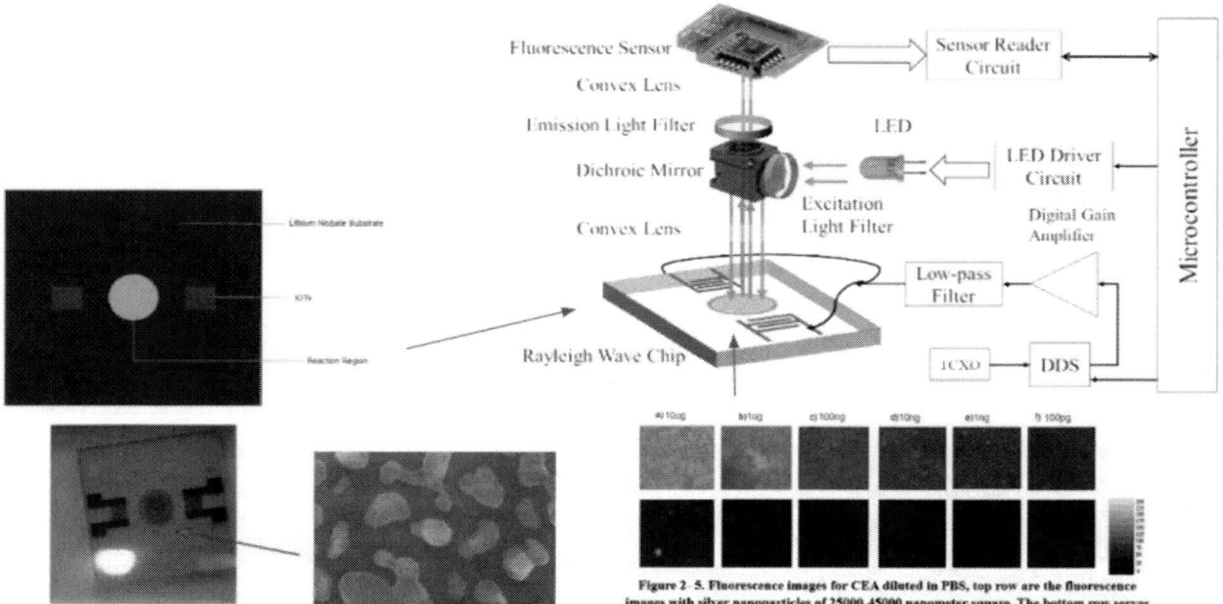

Figure 2-5. Fluorescence images for CEA diluted in PBS, top row are the fluorescence images with silver nanoparticles of 25000-45000 nanometer square. The bottom row serves as control groups without silver nanoparticles. The concentrations of CEA antigens in the images are (a) 10 μg/ mL, (b) 1 μg/mL, (c) 100 ng/mL, (d) 10 ng/mL, (e) 1 ng/mL, and (f) 100 pg/mL.

609302 High-Throughput Quantification of Influenza A Virus RNA Using Novel Drop-Based qRT-PCR Analysis

Abstract:

Influenza A virus (IAV) causes seasonal epidemics and sporadic worldwide pandemics that have resulted in over 100 million deaths during the past century. IAV populations are dynamic with high mutation rates and short replication times. A better understanding of IAV diversity at the single cell level would provide greater insight into how heterogeneity of infection affects population-level dynamics. Investigating IAV and host cell dynamics is typically performed using bulk culture and may not fully capture cell-to-cell differences in virus production. Thus, there is a need for high-throughput quantification of the IAV output, or burst size, from single cells to elucidate individual host-cell dynamics at the population scale. Here, we have developed a novel drop-based quantitative reverse transcriptase polymerase chain reaction (qRT-PCR) method to measure the amount of IAV RNA contained in individual drops. Concentrations of RNA are encapsulated in tens of thousands of picoliter-sized drops. These drops are processed on a microfluidic device which splits a small portion of the drop and merges it with qRT-PCR solution. The merged drops are thermocycled using a standard qPCR machine. Drops are randomly sampled at different cycles to measure drop fluorescence. Using a novel qPCR analysis method, we quantify the amount of RNA contained in the drops. Our method is the first to measure IAV RNA using a high-throughput drop-based assay, which will allow for future work in investigating the heterogeneity of IAV infection.

597969 Microfluidic Paper-Based Analytical Devices Using Plasma Processes

Abstract:

Disability adjusted life years (DALY) are the sum of years of life lost due to premature mortality and years lived with a disability. In 2010, DALY were approximately 2.4 billion. It is estimated that ~90% of DALY are due to communicable and non-communicable diseases. This problem is even more severe in developing countries where the major stumbling block for disease control is lack of proper disease diagnostic methods. Point-of care testing (POTC) methods in which users do not require specialized equipment and sophisticated lab facilities for disease diagnostics will be ideal for developing countries. Therefore, the World Health Organization (WHO) has stipulated a set of criteria for POCT in resource-limited settings called ASSURED criteria—affordability, sensitivity, specificity, user-friendly, rapid and robust, equipment-free and deliverable to those who need it. Due to the necessity and increasing demand of personalized medicine, the global POTC market is expected to grow from US $23.16 billion in 2016 to US $36.96 billion in 2021 at the compound annual growth rate of 9.8% from 2016 to 2021.

Microfluidic paper-based analytical devices (μ-PAD) have shown considerable promise to meet the ASSURED criteria set by the WHO. Paper has been patterned using various fabrication methods to allow distribution of liquid samples to targeted locations where subsequent reactions with reagents deposited in the paper are performed. These reactions develop color signals that can then be read using a smartphone. Low cost, portability, and the absence of external power source requirements make these devices desirable for POTC in less developed areas. It has been more than a decade since the first μ-PAD was invented. Yet these μ-PADs are not yet commercialized because of multiple technical problems including 1) contamination and evaporation of samples because of the lack of proper μ-PAD packaging, 2) inability to carry out complex tests that require multiple reagent additions, and 3) loss of sample in flow channels.

To address the issue of contamination and evaporation, a fully enclosed paper microfluidic device has been fabricated using pentafluoroethane (PFE) plasma deposition followed by O_2 plasma etching. Structures with one and two non-interacting, fully enclosed hydrophilic channels were generated in a single paper sheet using metal masks. Furthermore, by performing an additional O_2 plasma step with a secondary mask, pinholes were created at the reaction zones for reagent loading. Finally, to demonstrate the functionality of the device, a glucose assay was performed. Quantitative analysis of glucose assays showed that the device can be used for clinically relevant range of glucose. To our knowledge, this is the first time that such structures have been fabricated without paper stacking. Multi-layer devices have enhanced functionality relative to a single channel device, because the design space for creating networks of channels within the paper substrate is greatly expanded. The fluid-filled channels in the fabricated device are isolated, thereby preventing contamination due to handling and environmental exposure. Fluid evaporation can be inhibited by sealing the device with adhesive tape without contaminating the enclosed channels hence preventing loss of sample due to evaporation; a critical factor while working with precious samples.

To address the challenge of fluid flow control in μ-PADs, we extended our work on fully enclosed channel fabrication inside paper substrates to incorporate flow control functionality into the channels. Using pentafluoroethane plasma deposition and subsequent O_2 plasma etching with metal masks, fully enclosed hydrophilic channels with varying wettability were designed and fabricated to precisely control fluid flow in μ-PADs. Flow control is one of the most critical functionalities in μ-PADs because it allows controlled implementation of multiple sequential reactions on a single device with minimal user interference. Potential benefits of flow control include improved limit of detection, ability to carry out complex diagnostic tests, reduced assay time, and decreased likelihood of human error. We studied the effect of controlled fluorocarbon removal on the surface of cellulose fibers on channel wettability and demonstrated that this behavior can be exploited to significantly vary the fluid velocity in the flow channels. Furthermore, by simply covering the flow channels with adhesive tape, we demonstrated that the effect of evaporation on the fluid flow can be significantly decreased. Finally, a device with three channels converging to a single reaction zone was fabricated and sequential delivery of fluid to the reaction zone was demonstrated. The devices made by the methods described in this paper offer significant benefits for in-field diagnostic testing where multistep protocols are required.

A device fabricated by this method is estimated to cost between 10 and 50 ¢ and the plasma processes used to fabricate these devices are compatible with roll-to-roll technology, making it suitable for large scale production. This alternative fabrication approach contributes significantly toward commercialization of low-cost μ-PADs that meet the WHO ASSURED criteria.

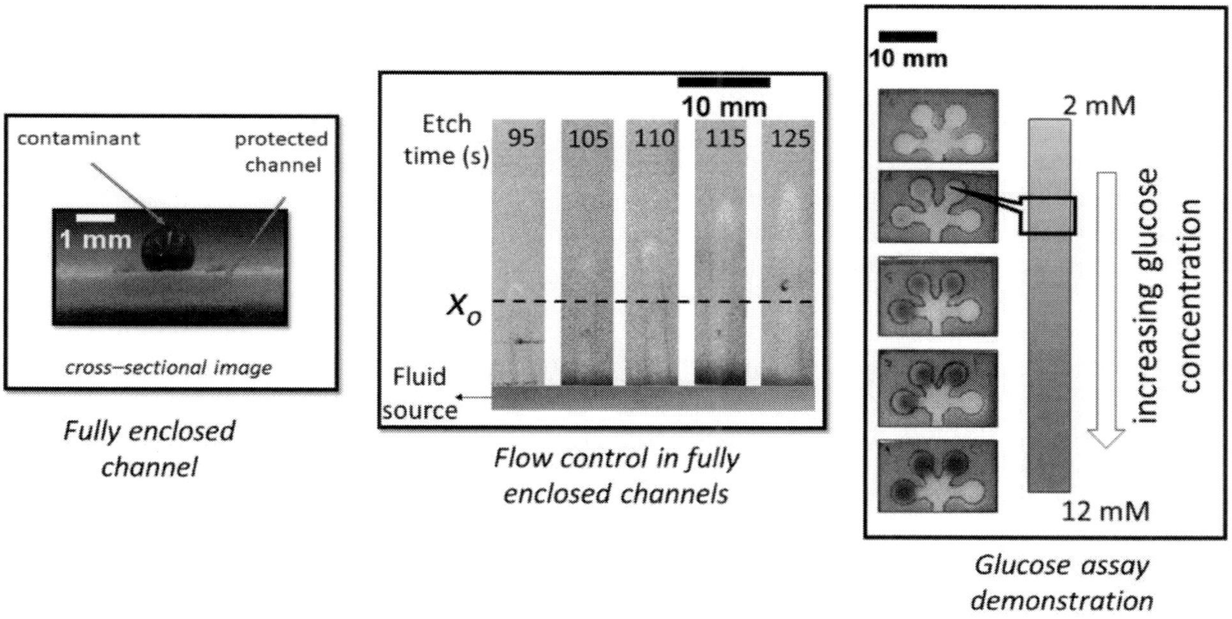

607198 Detection of Atrazine and Its Metabolites Using Photonic Molecularly Imprinted Polymers

Abstract:

Wetlands are beneficial in numerous ways and so, keeping them healthy and clean is an essential goal to fulfill. Due to use of herbicides for farming purposes, runoffs after raining would bring uncontrolled amounts of these herbicides to the wetlands. Atrazine is one of the most common herbicides used by farmers in the United States due to their selectivity against broadleaf weeds and annual grasses. Extensive use of this herbicide has caused contaminations in drinking water and soil and it can be harmful to living organisms even at low concentrations. According to EPA regulations, the drinking water maximum contaminant level (MCL) for atrazine is 3 ppb and being exposed to atrazine above this level may lead to lung, heart, and kidney diseases, low blood pressure, muscle spasms, weight loss, and damage to adrenal glands. Hence, it is important to monitor the level of atrazine in drinking water and ground water. Atrazine is degraded by plants and microorganism to metabolites which some of them are hazardous and detecting desethylatrazine (DEA) and desisopropylatrazine (DIA) is as essential as atrazine. The methods currently used to detect atrazine in different samples include gas chromatography (GC), gas chromatography-mass spectrophotometry (GC/MS), high pressure liquid chromatography (HPLC), liquid chromatography mass spectrophotometry (LC/MS) and enzyme-linked immunosorbent assay (ELISA). The gold standard for atrazine analysis, LC-MS, is precise but expensive, complex and time-consuming. The ELISA method, although relatively precise, has been found to be inaccurate in many cases and have positively biased results due to other organic materials interferences. Therefore, a method which is inexpensive, rapid and facile applicable is highly needed.

We developed an inexpensive label-free photonic sensor for the rapid detection of atrazine and its metabolites for field use. The sensor is composed of a molecularly imprinted polymeric (MIP) porous film that selectively captures atrazine molecules; the film pore structure is an inverse opal and its optical properties change when atrazine is captured.

Molecularly Imprinted Polymers (MIPs) have highly specific adsorption sites in their polymeric structure which leads them to selectively adsorb target molecules in those sites. Using a colloidal crystal of silica nanoparticles, a porous structure has been created in the polymer to facilitate target molecules movements through the thin film of polymer. By combining colloidal-crystal templating and molecular imprinting technique, a *p*H-paper-like sensor for a label-free colorimetric detection of atrazine and its metabolites in aqueous solutions and wetlands is fabricated. The void spaces of pores with different sizes and morphology generates Bragg diffraction and the wavelength of the peak changes when rebinding of target molecules happens. An optical signal via a change in Bragg diffraction is produced to detect atrazine. This optical property gives us an appropriate method of transducing recognition of target to a sensible signal.

Silica particle colloidal crystals were obtained by vertical self-assembly from 0.1% volume fraction suspensions in ethanol at 50°C. MIPs were fabricated with acrylic acid as the functional monomer, ethylene glycol dimethacrylate as crosslinker and UV-polymerization. The films, approximately 5 mm thin, were supported on polymethylmethacrylate (PMMA) slides.

Acrylic acid (AA), ethyleneglycol dimethacrylate (EGDMA), and azobisisobutyronitrile (AIBN) have been used as monomer, crosslinker, and initiator in polymerization precursor, respectively. Target molecules were also added to polymerization precure at 1:80 molar ratio to monomers. Then, a PMMA plastic slides on each sides of glass slides with depositions were firmly held together while immersed in polymerization precursor from one end. The polymerization precursor infiltrated through sandwich structure. Then, sandwich structure was held before UV light (365 nm) for 3 hours for the polymerization to take place. Inverse opal was created within polymer film by soaking PMMA slides in hydrofluoric acid (5%) and the silica particles were etched away, leaving empty cavities in the polymer. Finally, the target molecules were removed from film by washing with acetic acid and ethanol with ratio of 1:9. The MIPs were washed 3 times, for 15 minutes each time, using a shaker to enhance efficiency. Since detecting the metabolites is as essential as atrazine itself, MIPs were fabricated to detect DEA and DIA following the same steps.

Colloidal crystals were imaged by a FEI Quanta 600 FEG scanning electron microscopy to characterize the morphology of the crystal structure and the number of layers.

UV-Vis spectrophotometry was used to measure film surface reflectance to verify any swelling or shrinkage within the hydrogel. The reflectance spectra were recorded before and after exposure to clean water or atrazine over a wavelength range of 200–800 nm, using a benchtop double-beam UV–visible spectrophotometer equipped with a specular reflection accessory at a fixed angle of 30°.

Kinetic tests were conducted to investigate the time required for MIPs to reach equilibrium with a solution. Reflectance spectra were measured in specific time windows until the peak wavelength did not show any further changes. In kinetics test, the MIPs took 20 minutes to reach equilibrium, meaning it is necessary to incubate MIPs in unknown solutions for 20 minutes for the precise quantification of atrazine concentration.

The reflectance spectra of MIPs incubated in different solutions are illustrated in Figure 1. Changes in peak wavelength are a sign of swelling or shrinkage in hydrogel; in other words, adsorption of target molecules in binding sites of the polymer. The immersion of MIPs in solutions with higher concentration of atrazine changed the peak wavelength of the reflectance spectrum, as shown in Figure 1. In each step, peak wavelength increases in comparison with last more diluted solution which is a result of rebinding of target molecules in void binding sites in the structure of hydrogel. This target rebinding forms changes in porous structure of film, yielding different responses in reflectance spectrum each step. Then, based on this behavior and saturation time (20 minutes), MIPs were incubated in standard solutions with specific concentrations for 20 minutes with a gentle shaking and their reflectance were measure each step. The standard solutions, validated by LCMS, were used for measuring peak wavelength shifts for MIPs and to generate a calibration curve for quantifying unknown solution concentration of atrazine based on changes in wavelength shift.

The protocol for calibrating metabolites MIPs has been the same. Standard solutions validated by LCMS have been prepared and used to generate calibration curves by incubating metabolites MIPs in those standard solutions for 20 minutes on a shaker.

The fabricated sensor constitutes a simple, fast, and efficient tools to detect atrazine and metabolites in wetlands and also, quantify their concentration in order to make sure it does not go far higher than MCL. The thin film MIP-based sensors have a dynamic range of 0.14-12.3 ppb and calibration curves have been generated incubating MIPs in standard solutions with different concentrations.

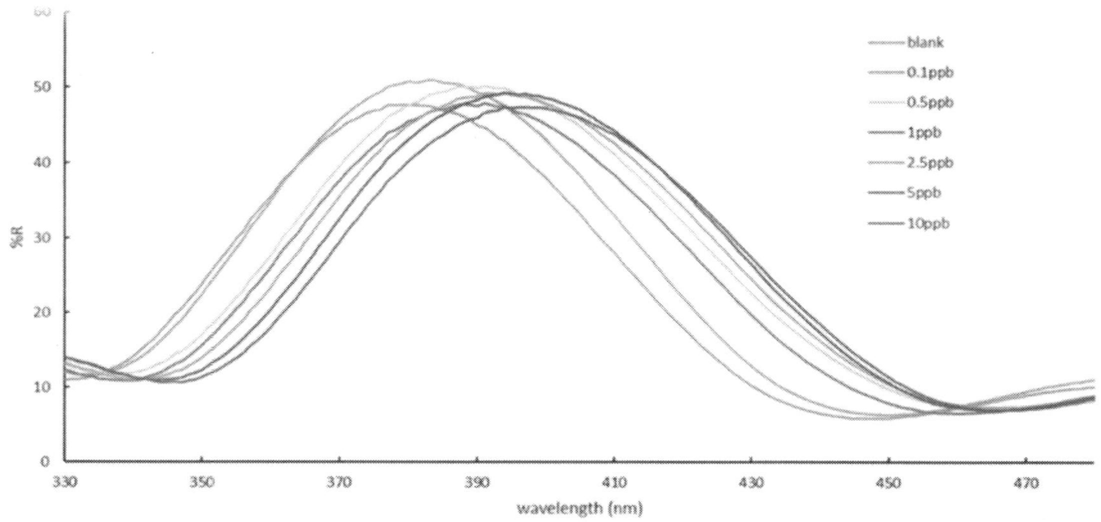

Figure 1. Optical responses of MIP to atrazine in a series of concentration.

598049 Simultaneous Detection of Harmful Herbicides with Lateral Flow Immunoassay Catalyzed By Palladium@Platinum Nanoparticles

Abstract:

In this presentation, we will show a method for simultaneously detecting two herbicides, atrazine and acetochlor, in real samples obtained from fruits and vegetables. These herbicides may be detected by lateral flow immunoassay (LFIA) using platinum coated over palladium (Pd@Pt) nanoparticles as a signal-amplifying agent. For validation, the results are compared with those obtained using a conventional HPLC analysis.

LFIA, also known as a test strip or an immunochromatographic assay, is an increasingly common immunoanalytical method for on-site screening because it is rapid, portable, inexpensive, and reagents are stable. LFIA has been applied in diverse areas such as medical diagnostics, drugs-of-abuse testing, environmental monitoring, and food safety. This study involves the use of LFIA for environmental monitoring and food safety, focusing especially on the detection of herbicides, since the capability of detecting these chemicals in the environment and in foods is crucial due to their potential hazards, from skin irritation to cancer. Atrazine is one of the most widely used herbicides in the United States, with 70 million pounds applied annually, as reported by the United State Geological Survey (USGS); however, the Environmental Protection Agency (EPA) has set the maximum contamination level in drinking water at 3 ng/mL for atrazine, and several studies have reported that atrazine has the potential to impact reproductive hormone levels in humans, rats, and amphibia. Acetochlor is another widely used herbicide, with an estimated 35 million pounds applied per year, as reported by the USGS; it is sometimes used instead of atrazine because of its superior biodegradability and higher maximum contamination level for drinking water, which is set at 20 ng/mL according to the Minnesota Department of Health. However, acetochlor is sited for its potential toxicity in rats and humans and has been noted by the EPA as having "Suggestive Evidence of Carcinogenic Potential."

Plans are underway to apply our previously developed dual-LFIA method, which employs a triangular sample pad combined with two test strips, to detect atrazine and acetochlor simultaneously in real samples of fruits and vegetables. We expect to be able to detect two herbicides near or below the maximum contamination levels set by the EPA and obtain recoveries between 80 – 120 %. The mesoporous Pd@Pt nanoparticle was used as the signal-amplifying agent because of its electrochemical activity as a peroxide conversion catalyst which allows for quantitative detection. Moreover, its high surface area increases its sensitivity, and its pH- and thermo-stability make it suitable for use with real samples. For validation, we detected the two herbicides simultaneously with HPLC and obtained a limit of detection of 10 ng/mL for atrazine and 50 ng/mL for acetochlor. We spiked apple and watermelon juices with 50, 80, and 100 ng/mL of atrazine and 100, 150, and 200 ng/mL of acetochlor, and obtained recoveries between 99.7 – 100 % when compared to an HPLC analysis. Laboratory experiments are planned for detecting the two herbicides in blended fruits, apples and grapes, and blended vegetables, corn and cabbage. We expect LODs near EPA limits and will report on the most recent progress. We expect our findings will suggest that innovations in efficiency, sensitivity, and stability are possible in tests for harmful chemicals in the environment and in food by use of the Pd@Pt nanoparticle enhanced immunoassays.

607001 Palladium Nanowires@Rgo@ZIF-8 Nanocomposite for High-Performance Hydrogen Sensors

Abstract:

Hydrogen forms a highly explosive mixture above 4% concentration in air. Therefore, fast detection of hydrogen in air is of high importance. Palladium has attracted attention as a material for room-temperature hydrogen sensing due to its unique ability to dissociate H_2 and form a hydride phase with higher electrical resistance than Pd metal under ambient conditions. However, the oxygen present in air often interferes with H_2 detection by generating moisture at the surface of Pd. Moreover, low-cost fabrication of Pd-nanowire sensors often produces sensors with non-monotonic response, due to competing effects of improved contact between nanowires and increased resistance of individual nanowires upon hydrogen uptake. Here, we report the fabrication of fast H_2 sensors using a novel ultra-thin and high-aspect-ratio Pd nanowires@reduced graphene oxide (PdNW@rGO) nanocomposite material coated with a layer of zeolite imidazole framework (ZIF-8). The rGO prevents direct contact between PdNWs and provides additional H_2-concentration dependent conduction channels, while the ZIF-8 layer limits oxygen transport to the NW surface. High-aspect-ratio Pd nanowires were synthesized in high yield by a hydrothermal method, followed by sonication with GO and subsequent reduction to form the PdNW@rGO nanocomposite. Pd nanowires of high-aspect-ratio increase the sensitivity of the sensor towards hydrogen while the layer of wrapped graphene prevents changes in NW contact resistance (i.e. creation of new conduction channels) due to volume expansion of PdNWs on hydrogen exposure. The ZIF-8 layer acts as a filter screening larger molecules like oxygen and nitrogen while allowing smaller molecules like H_2 to go through it. The hydrogen sensors on an interdigitated electrode (IDE) were exposed to concentrations (0.004-2%) of hydrogen diluted in synthetic air. The PdNWs@rGO@ZIF-8 hydrogen sensor showed a high-response of up to 4% resistance increase in air with a response time of 5 seconds to 1% H_2 in air and a lower limit of detection (LOD_{H2}) of 40 ppm. Relative to other fast Pd-based hydrogen sensors, the proposed solution-phase synthesis of PdNWs@rGO@ZIF-8 nanocomposite offers a combination of low cost and high-performance that is well-suited for applications such as fuel cell vehicles and hydrogen fueling stations.

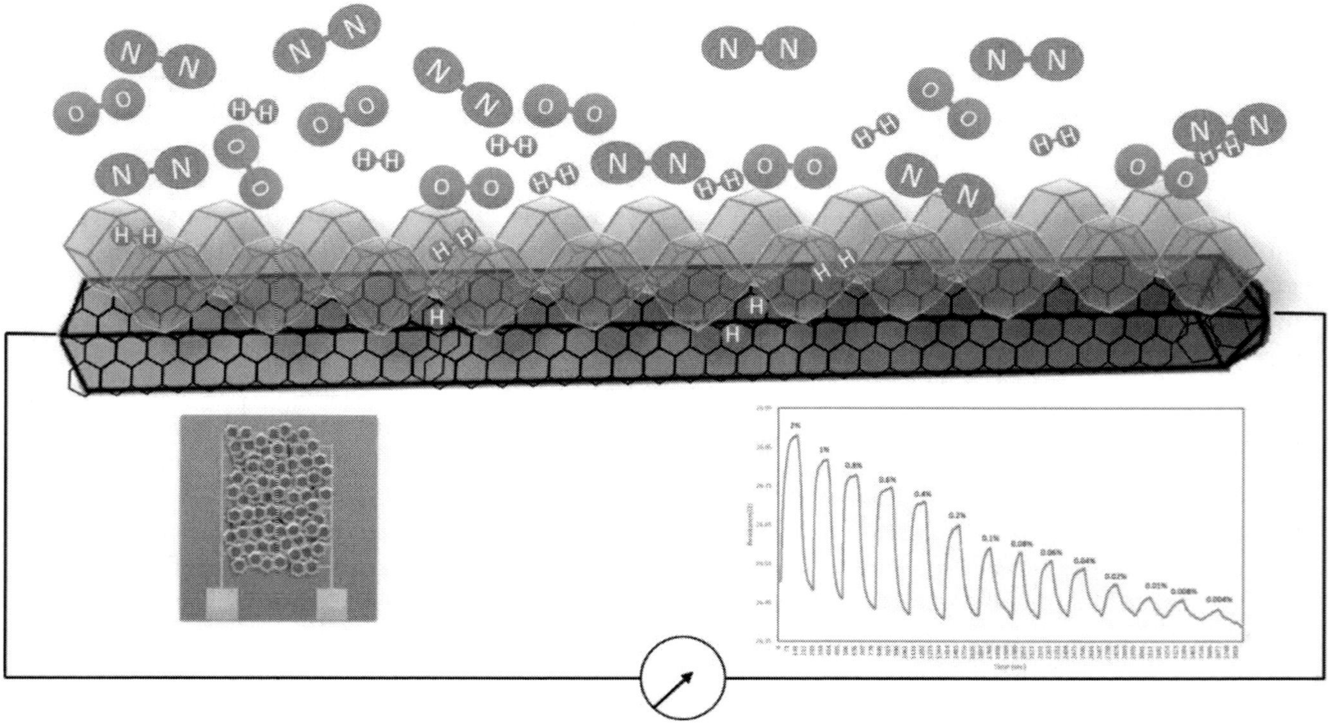

609139 Acoustic Wave Sensor Based on Plasticized Polymer Films for Detection of BTEX Compounds in Air

Abstract:

Although a variety of sensor technologies are available for environmental monitoring like photoionization, gas chromatography, mass spectroscopy, and e-noses sensors, chemical sensors based on piezoelectric transduction like the quartz crystal microbalance (QCM) and surface acoustic wave (SAW) device show promising results in terms of providing cost-effective real- time data, robustness, and low power requirements. Sensing films based on polymer-plasticizer films have been developed to detect the volatile organic compounds in the air, primarily benzene, ethylbenzene, toluene, and xylene (BTEX) using acoustic wave devices. This work shows for the first time how the selectivity and sensitivity of a glassy polymer film can be altered by adding a precise amount and type of plasticizer for BTEX detection.

To quantify the film saturation dynamics and understand the absorption of BTEX analyte molecules into the bulk of the sensing film, a diffusion study was incorporated to interpret the frequency vs time curve from a QCM. The sensor responses for individual analytes like benzene, toluene, etc. were collected and fit to exponentials which gives a characteristic value for response time, tau for individual analytes like benzene, toluene, ethylbenzene, and xylene which were found to be independent of the concentration of the analyte, thus helping in identifying the analyte. The model was also able to quantify individual analyte concentrations from the mixture based on the difference in tau values using a single polymer-plasticizer film on a single sensor. Several polymer/plasticizers materials and concentrations were screened to optimize the film for best BTEX sensing performance.

Quantifying BTEX concentrations from a mixture has always been challenging due to similarity in the chemical, physical, and molecular properties of the constituents. In this work, a single polymer-plasticizer film-based sensor has been employed to detect and quantify BTEX components in air based on sorption properties (solubility, partition coefficient, tau) as opposed to the traditional approach of using sensor arrays.

601716 Assessment of Residual Life of a Gas Filter Using Carbon-Nanotube Gas Sensors

Abstract:

A gas-filtration system filled with porous adsorbent is one of the most useful tools for reducing the amount of toxic volatile organic compounds released to the environment. However, it is challenging to assess the residual life of a filtration system and predict the time to replace the filter. In this study, we have shown how to evaluate the residual life of a gas-filtration system by embedding gas sensors into the system. Accordingly, chemiresistive gas sensors were fabricated via the deposition of patterned electrodes on a thin film of carbon nanotubes (CNTs). Multiple sensors were sequentially installed in the gas filter along the gas-flow direction. Dimethyl methylphosphonate, and ammonia were used to form a toxic gas model, and we observed the real-time changes in the resistance of the CNTs, which are highly sensitive to the adsorption of gas molecules because of their one-dimensional structure with high surface-to-volume ratio. The sensors responded sequentially as the toxic gases broke through the bed, and these responses could be correlated to the breakthrough curve of the toxic gases, obtained using gas chromatography and FT-IR. Here the properties of the CNT sensor can be potentially tuned by surface chemistry for the sensitive and selective detection of a target organic compound. Therefore, the approach reported herein can be useful for real-time assessment of the residual life of various gas-filtration systems.

601487 Ink Synthesis and Screen-Printing of Graphene Oxide Ink for Paper Based Electrochemical Sensors.

Abstract:

There has been a growing interest in making high performance electrode materials from graphene oxide (GO) to be used in electrochemical sensors, which has since gained popularity due to the excellent analytical performance, low-cost and simple properties linked to the devices. Herein, a method for fabricating highly conductive GO and GO@g-C3N4based ink to be used for screen printing electrochemical sensors. The physical properties of graphene oxide and GO@g-C3N4 were characterized by different measurements. Methods: Brunauer–Emmett–Teller (BET) surface area measurements, scanning electron microscopy (SEM), and Raman analysis, and GO was prepared using the modified Hummers' method. Furthermore, the viscosity, conductivity and surface tension of the printable GO and GO@g-C3N4 based ink were studied. Disposable screen-printed electrode electrochemical sensors were developed from GO-based inks for detecting heavy metals in effluent wastewater. Therefore it is expected that the GO-based inks will increase the sensitivity and selectivity of the electrode sensors.

605026 Effect of Surface Heterogeneity, Heat of Adsorption and Surface Area on the Characteristics of PANI-SnO_2 Based H_2s Gas Sensor

Abstract:

The emeraldine salt form of polyaniline (PANI) is a p-type organic semiconductor with polarons as charge carrier species[4] (figure 1 (a)). The protonation level of PANI changes upon exposure to a reducing or an oxidizing gas, which results as change in conductivity of the material[2]. Composites of the PANI with several metal oxides have been used to improve the sensor characteristics like sensitivity, selectivity, response time and limit of detection. SnO_2, one such metal oxide, makes a co-ordination with the N in PANI which results in a p-n junction between p-type polyaniline and n-type SnO_2[3]. Also, co-ordination with SnO_2 affects the protonated sites of polyaniline which introduces surface heterogeneity. The sensor characteristics are affected by surface heterogeneity as well as electrical properties of the material. Apart from this the surface area and morphology of the material also plays a major role in assisting the adsorption of the gas molecules on the sensor surface. In this work the effect of mixing the different amounts of SnO_2 in polyaniline on the H_2S sensing characteristics has been studied. The fractional change in electrical resistance of sensing element upon exposure to H_2S gas was recorded as the sensor response, and the surface site kinetics were calculated according to nature of the sensor response. The surface site kinetics were studied with Langmuir and Freundlich isotherms and the kinetics parameters were compared with the material properties such as heat of adsorption, surface area, and band gap.

PANI-SnO_2 composites were synthesized by solution casting method[3]. Four different wt. % (0, 5, 10 and 15) of SnO_2 nanoparticles (Sigma Aldrich, particle size < 100 nm) were added to 20 kDa molecular weight PANI emeraldine base (Sigma Aldrich) in 1 molar solution of camphorsulfonic acid (CSA) (Sigma Aldrich). The mixture was kept on stirring for 24 hours, during mixing the PANI gets protonated by CSA and SnO_2 forms co-ordination with N of PANI as shown in Figure 1 (b). After this reaction mixture was filtered, washed with DI and acetone, and dried. The composite powders were labeled as P-00, P-05, P-10, and P-15 according to increasing order of the SnO_2 loading.

The surface area of PANI decreased by addition of SnO_2, but the pore size was unaffected (table 1). The UV-Vis spectra showed π-π^* and polaron-π^* transitions at 350 nm and 450 nm respectively, and a free carrier tail which confirmed polaron form of polyaniline with expanded coil structure[5]. Band gap of the material was increased with 5 wt. % SnO_2 due to non-uniform dispersion of the n-type metal oxide. However, the band gap decreased on further addition of SnO_2 due to more co-ordination of Sn with PANI. The N1s XPS spectra confirmed amine, imine and N^+ type surface nitrogen moieties[6]. The percentage of active sites for each composite were approximated from N1s spectra.

The sensing elements were fabricated by drop casting the 10 wt. % slurry of composite powders in DI water over thermally deposited silver interdigitated electrodes on PET substrate. The sensor was placed in a 125 mL cubical chamber and nitrogen gas was flushed to make the atmosphere inert. Calculated amounts of H_2S gas were injected to make 10, 50 and 100 ppm concentrations inside the sensing chamber, and the electrical resistance of the sensing element was recorded by a digital multimeter (Keysight 34465a). The response time was calculated as the time taken by resistance in reaching the 63.2% of its saturation value. And the sensitivity was calculated in ppm^{-1} from the slope of response v/s H_2S concentration plot.

H_2S is a reducing gas and it also shows mild acidic behavior in presence of moisture[1], so it can protonate as well as deprotonate the PANI which result as change in the resistance. Since there was no moisture in sensing chamber, so the reducing behavior of H_2S caused deprotonation of the PANI, which increased resistance of the sensing element upon exposure to H_2S. For these sensors (P-00, P-05, and P-15) N^+ sites were considered as active site for chemisorption and surface site kinetics was evaluated from equation (1). However, the sensor P-10 had more uniform dispersion of SnO_2 in PANI, so there was more co-ordination of SnO_2 with PANI. In this case acidic behavior of H_2S caused protonation of PANI, and imine N sites were considered as active site of chemisorption. The chemisorption kinetics of surface sites was calculated from equation (2) for this sensor.

The response time of PANI sensor was improved on adding SnO_2 but the sensitivity was decreased (Figure 2(a)). Sensor P-10 show fastest response, but the sensitivity was decreased to 0.003 ppm^{-1}. Also, the reaction chemistry for P-10 sensor was different because the decreasing resistance response indicated protonation of PANI showing the acidic behavior of H_2S. The shorter response time for acidic behavior of the H_2S has been reported previously[1]. For other sensors (P-00, P-05, and P-15), the response time decreased with increase in band gap of the material, such behavior shows the importance of electrical properties of the material. The surface site kinetics (figure 2 (b)) show exponentially increasing behavior fitted with Langmuir kinetics (equation 3) and the adsorption isotherm show Freundlich (equation 4) behavior indicating heterogeneity in the surface sites of the sensing material. The adsorption coefficient α decreased with heat of adsorption for PANI-SnO_2 composites (figure 2 (c)). However, pristine PANI does not follow this trend due to more surface area (Table 1).

A high surface area with more active surface sites, and high heat of adsorption is desired for good sensitivity. A high band gap will give better response time of the sensor. However, the nature of surface chemistry will also play an important role in the sensor response characteristics. To design a PANI-SnO_2 composite based H_2S sensor, the environmental conditions (humidity level) of the sensor application area and synthesis procedure for more co-ordination of SnO_2 with PANI should be focused to promote role of the surface chemical reaction.

References

1. **Agbor, N.E., M.C. Petty, and A.P. Monkman**, "Polyaniline thin films for gas sensing," *Sensors and Actuators B: Chemical*, **28** (3), pp. 173–179 (1995).
2. **Hu, H., M. Trejo, M.. Nicho, J.. Saniger, and A. Garcıa-Valenzuela**, "Adsorption kinetics of optochemical NH3 gas sensing with semiconductor polyaniline films," *Sensors and Actuators B: Chemical*, **82** (1), pp. 14–23 (2002).
3. **Khuspe, G.D., S.T. Navale, M.A. Chougule, *et al.***, "Facile method of synthesis of polyaniline-SnO2 hybrid nanocomposites: Microstructural, optical and electrical transport properties," *Synthetic Metals*, **178**, pp. 1–9 (2013).
4. **Stafström, S., J.L. Brédas, A.J. Epstein, *et al.***, "Polaron lattice in highly conducting polyaniline: Theoretical and optical studies," *Physical Review Letters*, **59** (13), pp. 1464–1467 (1987).
5. **Xia, Y., J.M. Wiesinger, A.G. MacDiarmid, and A.J. Epstein**, "Camphorsulfonic Acid Fully Doped Polyaniline Emeraldine Salt: Conformations in Different Solvents Studied by an Ultraviolet/Visible/Near-Infrared Spectroscopic Method," *Chemistry of Materials*, **7** (3), pp. 443–445 (1995).
6. **Xiong, P., H. Huang, and X. Wang**, "Design and synthesis of ternary cobalt ferrite/graphene/polyaniline hierarchical nanocomposites for high-performance supercapacitors," *Journal of Power Sources*, **245**, pp. 937–946 (2014).

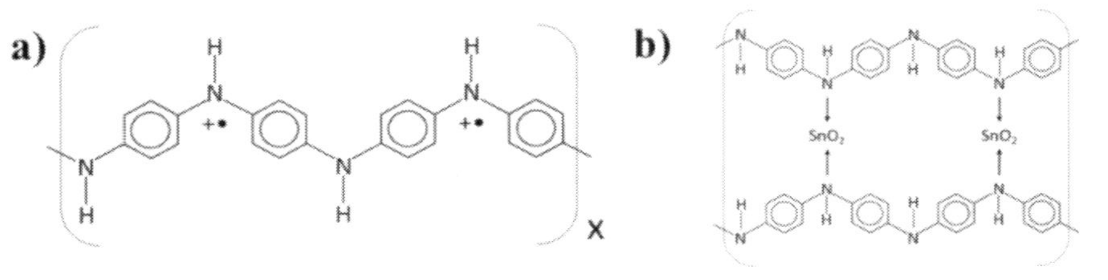

Figure 1. a) Polaron structure of Polyaniline Emeraldine Salt, b) SnO_2 co-ordination with PANI

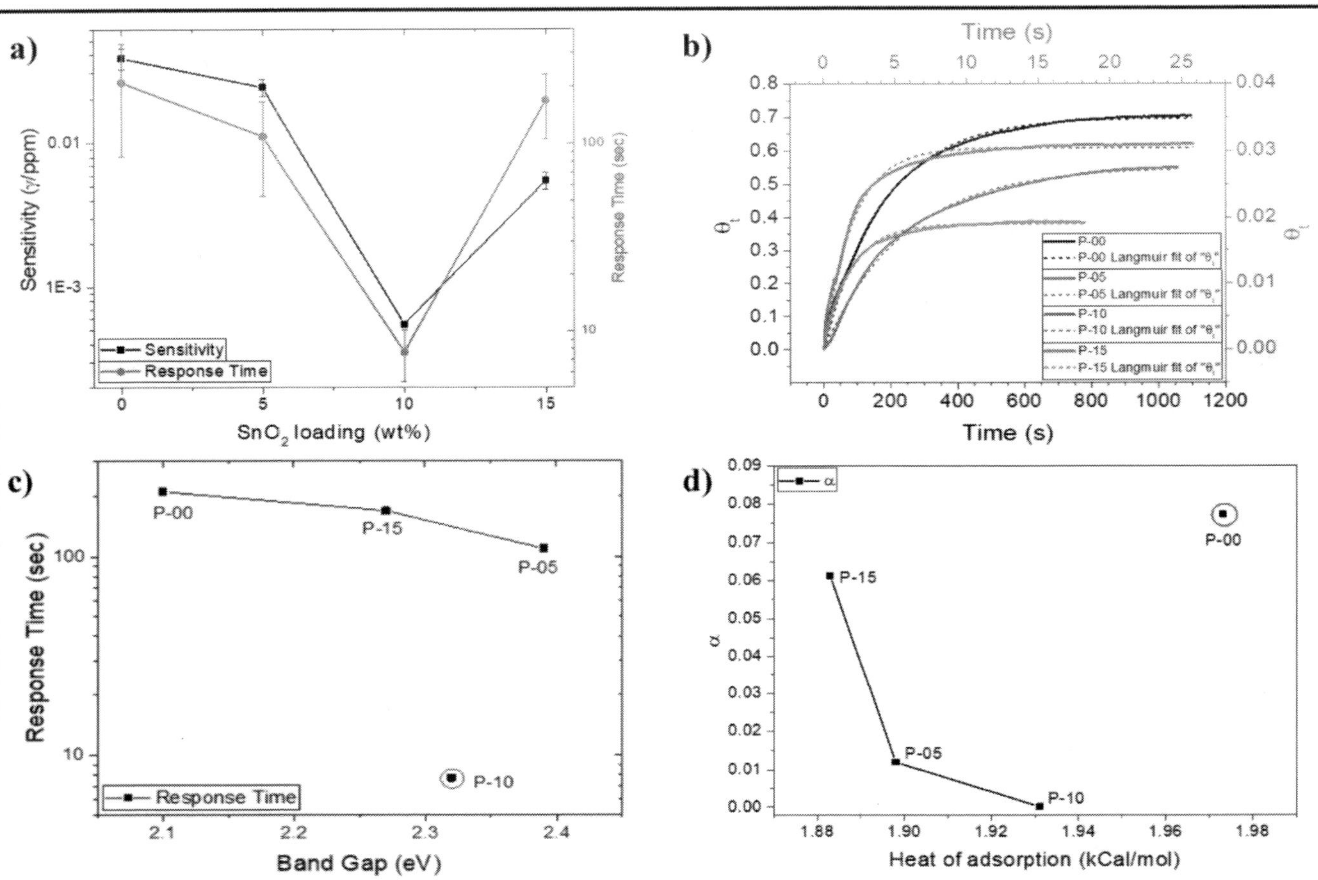

Figure 2. a) Sensitivity and response time of PANI-SnO_2 based H_2S sensors. b) Surface site kinetics for 50 ppm H_2S environment. c) Freundlich isotherm coefficient (α) and heat of adsorption (Q) correlation

Equations:

$$\theta(t) = \frac{\gamma(t)}{1+\gamma(t)} \quad (1)$$

$$\theta(t) = \frac{\gamma(t)}{1-\gamma(t)} \quad (2)$$

$$\theta(t) = \frac{N^*}{N_t} = \theta_0\{1 - \exp(-\nu t)\} = \frac{P}{P+b}\{1 - \exp(-\nu t)\} \quad (3)$$

$$\theta_0 = \propto P^{1/n} \quad (4)$$

Table 1: BET Surface area, heat of adsorption, and BJH pore size of the PANI-SnO_2 Composites

Material	BET surface Area (m^2/g)	Heat of adsorption (kCal/mol)	BJH Pore size (Å)
P-00	63.53	1.97	14.2
P-05	27.95	1.89	14.2
P-10	28.71	1.93	14.2
P-15	29.52	1.88	16.7

607925 Effects of Alkane Dielectrics in Chemically-Sensitive Field-Effect Transistors Functionalized with Metal-Organic Frameworks

Abstract:

There is a growing demand for distributed gas-sensing networks with sensors that have ultra-low-power consumption, high sensitivity, and selectivity. One such device, the chemically sensitive field-effect transistor (CS-FET), meets these criteria(1). The CS-FET is structurally similar to a standard field-effect transistor, except the gate metal is replaced by a sensing layer whose work function shifts when gas molecules adsorb. However, the CS-FET is susceptible to baseline drift from humidity adsorption at the hydrophilic SiO_2 gate dielectric, which can cause a 50% increase in source-drain current at just 40% relative humidity. A separate issue is that standard sensing materials are non-selective metals. Development of sensing materials that are selective and do not suffer humidity drift is needed to improve the performance of the CS-FET devices.

Towards addressing these problems, we have investigated the potential of metal-organic framework (MOF) sensing layer in CS-FET devices. MOFs are an exciting class of materials for chemical sensing because the adsorbate-driven work-function-shift is highly analyte-MOF specific. Furthermore, we have functionalized the SiO_2 dielectric of the CS-FET with hydrophobic self-assembled monolayers octadecyl-trichloro-silane (OTS) and dimethyl-dichloro-silane (DDMS). These monolayers can be subsequently coated with MOF sensing layers; in our case, HKUST-1 is used as the chemically sensitive layer for humidity because HKUST-1 is known to have a strong affinity for water(2).

The CS-FET devices are functionalized with DDMS or OTS. Sensors with just OTS have nearly no response to humidity, with about less than a 5% increase in source-drain current at 40% humidity. Sensors with just DDMS have a 10% increase in current starting at 40% humidity. Once these SAM-modified devices are coated with an HKUST-1 sensing film, we find distinct sensing dynamics and, surprisingly, different sensitivities. With OTS and HKUST-1, the device responds within seconds and is quite sensitive, achieving nearly 800% initial current at 40% relative humidity compared to dry air. However, with DDMS, the response is much slower and much less sensitive, with only 2x initial current at 40% relative humidity. We attribute the better dynamics to the better coverage of OTS compared to DDMS on the SiO_2. The difference in sensitivity is due to a band bending effect induced by the monolayer in the underlying silicon channel. Band bending is confirmed by differences in the electrical characteristics of the devices.

Inspired by these results we have begun to work on sensors that replace the SiO_2 dielectric completely with an alkane dielectric, 1-octadecene, that grows directly on silicon. Preliminary results show that a device with just this monolayer has even a weaker response to humidity than OTS. We plan to evaluate this monolayer as a dielectric for an HKUST-1-based humidity sensor and evaluate the effect of the monolayer on the electrical properties of the CS-FETs.

References:

1. H. M. Fahad, N. Gupta, R. Han, S. B. Desai, A. Javey, Highly Sensitive Bulk Silicon Chemical Sensors with Sub-5 Nanometer Thin Charge Inversion Layers. *ACS Nano*. **12**, 2948–2954 (2018).
2. D. W. Gardner, X. Gao, H. M. Fahad, A.-T. Yang, S. He, A. Javey, C. Carraro, R. Maboudian, Transistor-based work function measurement of metal-organic frameworks for ultra-low-power, rationally designed chemical sensors. *Chem. – A Eur. J.* (2019), doi:10.1002/chem.201902483.

597965 First Principles Modeling of Interfacial Charge Transfer and Redox Chemistry in Solid State Batteries

Abstract:

Solid electrolytes in rechargeable lithium ion batteries provide enhanced thermal stability over more traditional liquid electrolytes. Solid electrolytes could also enable the safe application of lithium metal anodes and in turn, an expanded voltage window in comparison to conventional lithium ion systems.[1] In spite of recent advancements in solid state battery technology with respect to Li^+ ionic conductivity, there remain many issues broadly associated with resistivity at the interface between electrodes and the solid electrolyte. For example, the electrochemical stability window is often limited by the oxidation and/or reduction potentials of solid electrolytes against high voltage cathodes and lithium metal anodes, respectively.[2] Under such oxidizing and reducing conditions solid-state superionic conductors are not thermodynamically stable, and various decomposition products are thermodynamically favored instead.[3]

In this work, we probe the atomistic details of such interfacial decomposition reactions by analyzing the well-documented reactivity[4] between lithium lanthanum titanate (LLTO) solid electrolytes and lithium metal anodes. First principles density functional theory (DFT) calculations are performed within a Li grand potential ensemble[2,3,5] to describe bulk thermodynamic driving forces. These thermodynamic driving forces are then mapped onto a La-Ti-O convex hull, where we subsequently generate explicit solid-solid interfacial models that likely exist under operating conditions. A subset of these interfacial models is then used to initialize *ab initio* molecular dynamics (AIMD) simulations. These AIMD simulations directly suggest structure-sensitive reconstruction mechanisms and redox chemistry based on the particular LLTO surface termination that is interfaced with the lithium metal anode. The atomistic insights from these AIMD studies combined with interfacial band alignment analysis[6] further unravels the physical origins of charge transfer[7] and redox chemistry at Li/LLTO interfaces. We apply these insights to propose synthetic strategies to kinetically stabilize electrode-electrolyte interfaces in solid-state batteries. These theoretical predictions are further supported by *operando* X-ray scattering studies. This overall approach combines DFT based bulk and interfacial thermodynamic analyses, AIMD simulations of explicit solid-solid interfaces, and band alignment calculations to reveal new design principles to enhance interfacial stability in solid state batteries.

[1] G. Crabtree, *Science* **2019**, *366*, 422.

[2] W. D. Richards, L. J. Miara, Y. Wang, J. C. Kim, G. Ceder, *Chem. Mater.* **2016**, *28*, 266.

[3] Y. Zhu, X. He, Y. Mo, *ACS Appl. Mater. Interfaces* **2015**, *7*, 23685.

[4] S. Wenzel, T. Leichtweiss, D. Krüger, J. Sann, J. Janek, *Solid State Ion.* **2015**, *278*, 98.

[5] R. E. Warburton, H. Iddir, L. A. Curtiss, J. Greeley, *ACS Appl. Mater. Interfaces* **2016**, *8*, 11108.

[6] K. L. Bassett, R. E. Warburton, S. Deshpande, T. T. Fister, K. Ta, J. L. Esbenshade, A. Kinaci, M. K. Y. Chan, K. M. Wiaderek, K. W. Chapman, J. P. Greeley, A. A. Gewirth, *Adv. Mater. Interfaces* **2019**, *6*, 1801923.

[7] M. W. Swift, Y. Qi, *Phys. Rev. Lett.* **2019**, *122*, 167701.

600842 Meso-Scale Architecture, Environment and Active Site: Evaluating Catalysts for Electrochemical CO_2/CO Reduction

Abstract:

High energy fuels supply 80% of global energy, combined with cement processes, this accounts for emission of 30 billion tons of CO_2 annually.[1–3] This non-renewable source of energy is not only tethered to the global economy, but each day we rely on it we jeopardize the global climate. Implementing technologies that sequester atmospheric CO_2 and close the carbon cycle are paramount for stability of both the economy and the global climate. The electrochemical reduction of CO_2 (CO_2RR) is one promising technology to help solve this problem. However, the process of electrochemically reducing CO_2 is challenging, as driving selectivity to a given hydrocarbon product and mitigating the parasitic hydrogen evolution reaction (HER) both impede progress.[4] In the pursuit of more active and selective catalytic materials, researchers have been moving towards high surface area nanocatalysts and away from the planer Cu surfaces studied initially.[5] This transition results in a complex meso-scale catalyst layer that complicates diffusion of CO_2 to active sites. Evaluation of the intrinsic activity of new materials for CO_2RR is moot if CO_2 is not sufficiently being supplied to a catalyst's surface, especially when the electrolyte provides a near infinite supply of H_2O for HER.[6] In this report I will discuss my exploration into CO_2 electrolysis, from the development of highly active Cu NWs that displayed promising activity at low overpotentials only to disappoint at high overpotentials.[7] We later created a model of the electrolyte conditions within these nanowires and found supporting evidence that depletion of CO_2 at modest potentials ($\approx$-0.6 V vs RHE) was the likely cause of the apparent catalyst activity.[6] Using this knowledge, I pursued alternative hierarchical Cu nanostructures that successfully demonstrated increased CO_2 consumption rates due to their lower active site density.[8] Having learned the impact of meso-scale architecture and CO_2 diffusion, the nanocatalysts were adapted to a gas-diffusion electrolyzer, moving the bulk of CO_2 diffusion from liquid-phase ($\approx 2x10^{-9}$ m^2/s) to gas-phase ($\approx 1.6x10^{-5}$ m^2/s) to enable higher CO_2 flux. In this configuration we have been able to investigate physical structure-function relationships concerning catalyst layer thickness and utilization of catalyst active sites. However, findings demonstrate that the gas-diffusion electrolyzer can give a false sense of security about mass transport, as the catalyst layer thickness continued to play a major role in determining the output of the electrolyzer. Even under high CO_2 flux conditions there is an interplay between selectivity (Figure 1d), activity (Figure 1e) and catalyst layer thickness. The culmination of this work has most likely raised more questions than it has answered about ideal conditions in a CO_2 electrolyzer, but my future endeavors will continue pushing the limits of what is known, ultimately contributing to viable technological solutions that end reliance on non-renewable energy sources.

References

(1) Guo, Z.; Liu, B.; Zhang, Q.; Deng, W.; Wang, Y.; Yang, Y. Recent Advances in Heterogeneous Selective Oxidation Catalysis for Sustainable Chemistry. *Chem. Soc. Rev.* **2014**, *43* (10), 3480–3524. https://doi.org/10.1039/C3CS60282F.

(2) Catalysts: - Specialty Chemicals Update Program (SCUP) | IHS Markit https://ihsmarkit.com/products/chemical-catalysts-petroleum-and-chemical-scup.html (accessed Apr 24, 2020).

(3) Global Carbon Project (GCP) https://www.globalcarbonproject.org/carbonbudget/ (accessed Apr 24, 2020).

(4) Verma, S.; Kim, B.; Jhong, H.-R. "Molly"; Ma, S.; Kenis, P. J. A. A Gross-Margin Model for Defining Technoeconomic Benchmarks in the Electroreduction of CO2. *ChemSusChem* **2016**, *9* (15), 1972–1979. https://doi.org/10.1002/cssc.201600394.

(5) Hori, Y.; Murata, A.; Takahashi, R. Formation of Hydrocarbons in the Electrochemical Reduction of Carbon-Dioxide at a Copper Electrode in Aqueous-Solution. *J Chem Soc Farad T 1* **1989**, *85*, 2309–2326. https://doi.org/10.1039/F19898502309.

(6) Raciti, D.; Mao, M.; Wang, C. Mass Transport Modelling for the Electroreduction of CO2 on Cu Nanowires. *Nanotechnology* **2018**, *29* (4), 044001. https://doi.org/10.1088/1361-6528/aa9bd7.

(7) Raciti, D.; Livi, K. J.; Wang, C. Highly Dense Cu Nanowires for Low-Overpotential CO2 Reduction. *Nano Lett.* **2015**, *15* (10), 6829–6835. https://doi.org/10.1021/acs.nanolett.5b03298.

(8) Raciti, D.; Wang, Y.; Park, J. H.; Wang, C. Three-Dimensional Hierarchical Copper-Based Nanostructures as Advanced Electrocatalysts for CO2 Reduction. *ACS Appl. Energy Mater.* **2018**, *1* (6), 2392–2398. https://doi.org/10.1021/acsaem.8b00356.

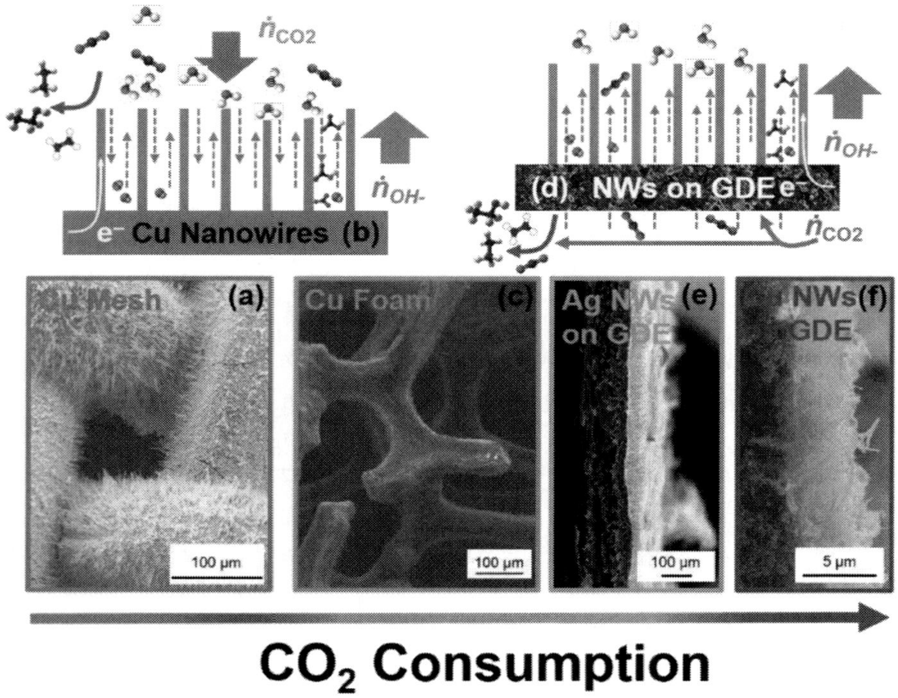

612165 Highly Efficient and Stable Unitized Regenerative Fuel Cells (URFCs) for Long-Term Energy Storage and Conversion

Abstract:

Global energy demand and consumption have skyrocketed since the last few decades. Environmental concerns such as climate change, air pollution and greenhouse gas emission, have limited the further overly usage of fossil fuels and emphasized the importance of deploying renewable sources. Therefore, reliable and efficient energy storage technologies are in urgent need to accommodate the growing capacity of intermittent solar and wind based power resources.

The unitized regenerative fuel cell (URFC) has been proposed as a promising electrochemical device for intermittent renewable energies storage and conversion. However, the widespread application has been hindered due to low round trip efficiencies (RTEs) and disappointing durability, in particular at high current densities. One of the main challenges to bring the URFC into practical applications comes from the employment of bi-functional electrodes for URFCs operation due to the combination of fuel cell and electrolyzer in one unitized device, which could bring constraints in designing electrodes to efficiently support half reactions for both fuel cell and electrolyzer. Therefore, it is urgent to develop high active and stable bi-functional electrode to support the efficient and durable operation URFCs.

Here, we demonstrate a design of bi-functional electrode by using commercially available Pt and Ir black catalyst for PEM-URFCs that significantly improved round-trip efficiency and durability of PEM-URFCs compared to the literature. An optimized Pt&Ir black electrode could achieve 56% and 53% of RTEs under CE and CG mode of operation at 1 A cm^{-2}, respectively, and continuous operation with negligible degradation under both CE and CG mode for over 500 h (Figure 1). The results demonstrate the viability of applying URFCs for long-term energy storage at previously unattainable efficiencies, and cast new light on electrode design and optimization of URFCs.

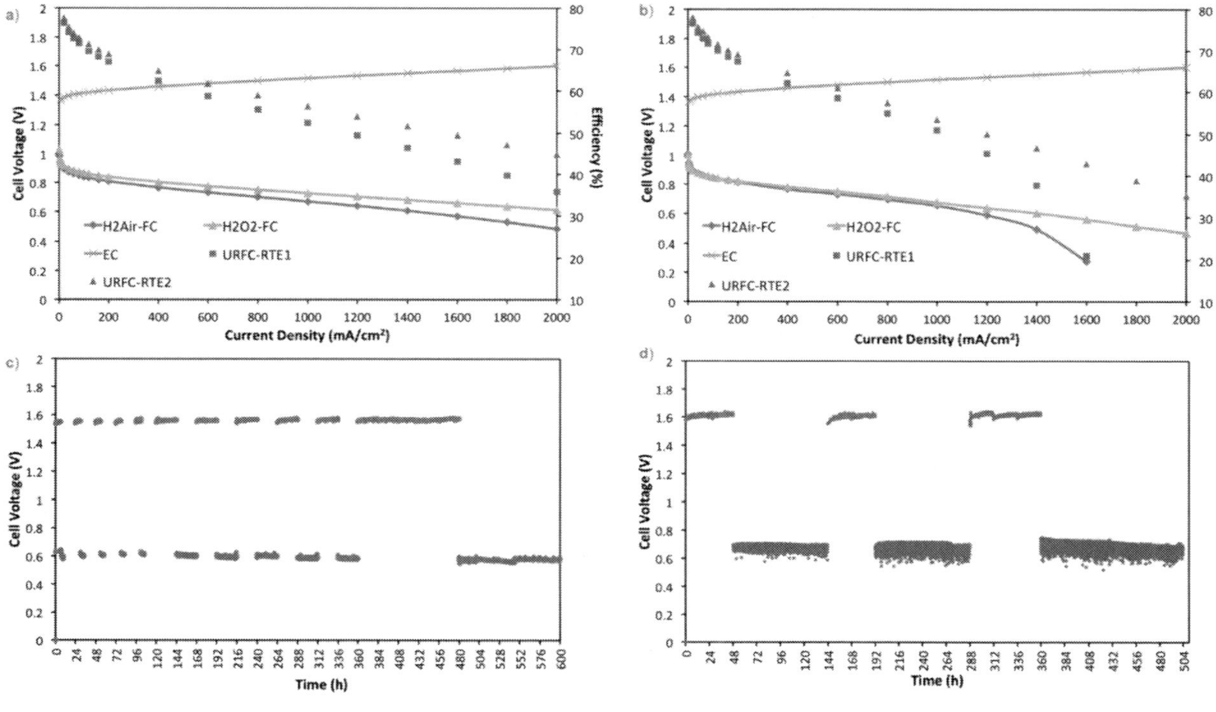

594236 Electrochemically-Mediated Carbon Dioxide Separation Using Redox-Active Molecular Sorbents and Process Intensification

Abstract:

Anthropogenic carbon dioxide emissions present a serious challenge to our society. One of the foremost mitigation strategies involves carbon capture, particularly from large stationary emission sources, followed by sequestration or utilization. The incumbent technology for carbon dioxide separation is temperature-swing-based amine scrubbing, which can be challenged by the substantial energy demand, sorbent degradation, environmental concerns and large footprint. Electrochemically-mediated separations offer a low-temperature, ambient-pressure alternative for carbon capture, representing a promising yet largely unchartered research area. By being electrically driven, these systems can be controlled precisely to reduce energy losses, are modular and thus easy to implement, and possess great adaptability to the multi-scale nature of carbon capture.

In this talk, an electrochemically-mediated carbon capture technique will be introduced, relying on redox-active quinoid species that undergo significant changes in their carbon dioxide binding affinity as they progress through an electrochemical cycle. Though the observation of reversible carbon dioxide binding by quinones has been made before, such chemistry can only be operated using flammable, toxic aprotic solvent, posing a great barrier toward large-scale implementation.

Here, we demonstrate that rationally designed aqueous electrolytes with high salt concentration can effectively resolve the incompatibility between aqueous environments and quinone electrochemistry for carbon capture. The salt-concentrated aqueous media offer distinct advantages including extended electrochemical window, high carbon dioxide activity, significantly reduced evaporative loss and material dissolution, and importantly, greatly suppressed competing reactions as verified by both experiments and theoretical calculations. When tested using simulated flue gas, our proposed system achieved outstanding performance metrics (capacity, stability, efficiency and electrokinetics) and competitive system energetics, advancing electrochemical carbon separation further towards practical applications. Moreover, through integration with a novel electrochemically-controlled gas-gating membrane that dynamically controls gas passage, an effectively continuous operation of carbon capture and release can be achieved, bringing new opportunities for process intensification.

604688 Applications of Advanced Fibers for Next-Generation Lithium Batteries: From Liquid to Solid-State Cells

Abstract:

Rechargeable lithium-ion batteries (LIBs) have received tremendous attention over the past three decades due to their relatively higher energy density, longer cycle life, and lower environmental impact compared to the traditional Ni-Cd/MH battery systems. Due to the continuously increasing demand of the world energy consumption, the general goal for battery development is to increase the energy densities. However, it is extremely difficult to improve the state-of-the-art LIBs' energy densities furthermore because of the limited capacities of the graphite anode (~370 mAh g^{-1}) and insertion-oxide cathode (~250 mAh g^{-1}). Therefore, alternative cathode and anode materials offering higher capacities should be developed.

Lithium-sulfur (Li-S) battery is one of the prospective candidates in this regard not only because both sulfur and lithium have high theoretical capacities but also because sulfur is low cost and environmentally friendly. Moreover, lithium metal has a low negative potential. Nevertheless, the commercialization of Li-S batteries has been hindered by several issues, such as the insulating nature of sulfur and its intermediates, polysulfide shuttle effects, safety concerns of the lithium metal electrode, etc. Intensive efforts have been taken to solve these problems by developing novel sulfur host materials, solid-state electrolytes, separators and interlayers, and binders, etc., during the past decade. It should be noted that advanced fibers have played extremely important roles in the abovementioned solutions due to their substantial surface-to-volume ratio, flexibility in surface functionalities, large porosity, and good mechanical properties.

Here, I will mainly concentrate my efforts on the advancements and progress of fibers employed for achieving high-performance Li-S batteries: from liquid to solid-state cells. The fundamental electrochemistry and challenges of Li-S batteries will firstly be discussed. It will then focus on advantages of using fibers in Li-S batteries and the mechanisms behind their operations which are detailed from two perspectives: i) separator selection and design; and ii) solid-state electrolytes. I will talk about how fiber-based separators improve the rate capability of Li-S cells, and how nanofiber-based solid-state electrolytes prevent the growth of the lithium dendrites, addressing the safety issue of Li-S cells. Finally, the critical research directions needed and the remaining challenges to be addressed in this area will be summarized.

606268 Transition Metal Carbide Electrodes Modified By High-Powered Impulse Magnetron Sputtering

Abstract:

Introduction

Transition metal carbides (TMCs) play a critical role in the functionality of many engineered systems related to energy storage, as well as applications requiring corrosion resistance, high temperature and high strength. It is well known that the physics regulating the material properties of TMCs is controlled by its structure and chemistry at both the atomic and mesoscales. Indeed, unconventional carbide phases published in the last few years have shown the ability to access enhanced electronic conductivity[1-2], and catalytic activity[2-3]. Transition metal carbides are typically synthesized under thermal equilibrium conditions using furnaces often operating in excess of 1500 °C. By contrast, the non-equilibrium[4] nature of low- temperature plasmas such as in high-powered impulse magnetron sputtering (HiPIMS) makes this technique an excellent candidate to synthesize novel non-equilibrium TMCs with unique material properties.

Our study of non-equilibrium TMC thin films generated by HiPIMS is characterized using microscopy, electrochemical corrosion, and electrochemical impedance spectroscopy in an effort to build empirical electrochemical-material relationships.

Materials and Methods

Molybdenum carbide (Mo_2C) micro particles synthesized by co-reduction carburization were immobilized on carbon paper thus mimicking a catalyst support dispersed on a gas diffusion layer (GDL). Thin films of deposited or implanted Mo, Ta, and Zr were grown using a range of kick pulses up to +700 V. Pt catalyst loading on the particles was 20 wt% and at 57 $\mu g/cm^2$). Diffractometry and spectroscopic techniques were used to determine material composition. Equilibrium corrosion curves (potential scan rate 0.167 mV/s) were acquired in high and low pH media. Chronopotentiometric titration (constant current, varying pH) was performed to determine passivation and dissolution currents similar to those produced in a Pourbaix diagram[5]. Electrochemical impedance spectroscopy (EIS) was used to determine resistance to an oxygen exchange reaction.

Results and Discussion

The threshold for improved performance was determined to be as low as 5 nm of Ta deposited/implanted using a kick pulse of 500-700 V. Mo_2C receiving this treatment displayed an increased corrosion potential of 100-110 mV compared to untreated Mo_2C, See Figure 1. Bulk crystallographic phase of the Mo_2C was unchanged, all alloys formed were in the ~ 5 nm closest to the surface leaving overall electrical conductivity unchanged.

Significance

With the treatment described above, a fuel cell utilizing Mo_2C catalyst supports would be able to run at approximately 20% higher average power without negative repercussions on device performance or lifetime.

Figure 1. Equilibrium corrosion curves for Mo_2C/C supports with and without the HiPIMS Ta treatment

References

1. Naguib, M.; Halim, J.; Lu, J.; Cook, K.M.; Hultman, L.; Gogotsi, Y.; Barsoum, M.W. *Am. Chem. Soc.* 135, 43 (2013).
2. Tang, C.; Zhang, H.; Xu, K.; Zhang, Q.; Liu, J.; He, C.; Fan, L.; Asefa, T. *Mater. Chem. A.* 7, 30 (2019).
3. Schmuecker, S.M.; Clouser, D.; Kraus, T.J.; Leonard, B.M. *Dalton Trans.* 46, 39 (2017).
4. Wang, Z.; Zhang, Y.; Neyts, E.C.; Cao, X.; Zhang, X.; Jang, B.W L.; Liu, C.-j. *ACS Catal.* 8, 3 (2018).
5. Weidman, M.C.; Esposito, D.V.; Hsu, Y.C.; Chen, J.G. *Power Sources.* 202 (2012).

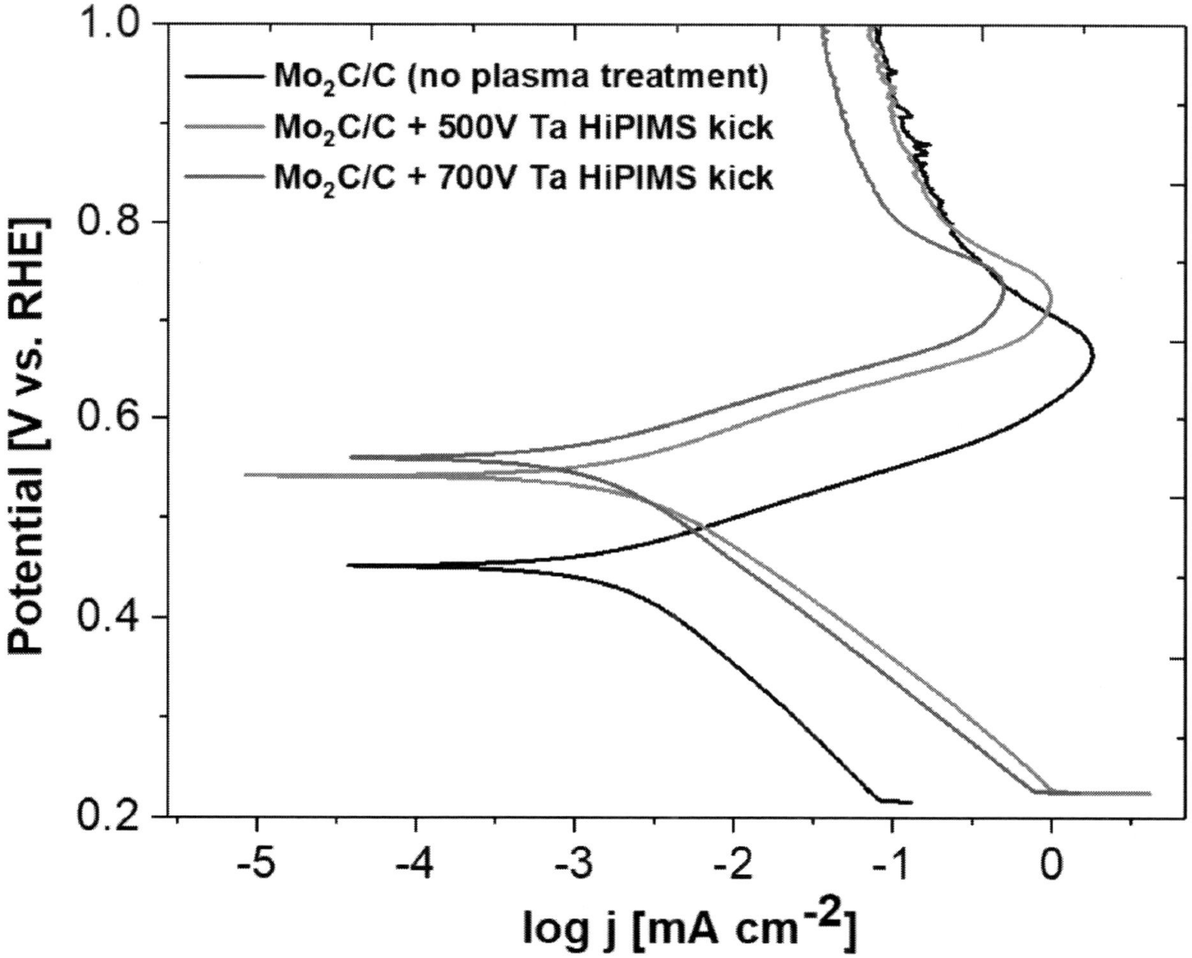

600204 Zinc Anode Design for High-Energy Rechargeable Aqueous Zn-Air Batteries

Abstract:

Alternative energy carriers have been sought after due to fossil fuels' slow regeneration and environmental concerns such as climate change, air and water pollution. When carrying clean electricity from solar or wind, batteries are promising to alleviate the current energy and environmental problems. Lithium-ion batteries are widely used energy storage system because of their outstanding energy density and rechargeability. However, safety is always a concern because the use of flammable organic electrolytes. On the other hand, batteries that use aqueous electrolytes have enhanced safety (**Fig. 1a**), ion conductivity, and cost-effectiveness. Within the stability window of water, zinc is an attractive anode material because it is the most active metal that is stable with water. By using aqueous electrolyte, zinc-based batteries not only are safer, but also can be manufactured in ambient air rather than dry room, and has much higher tolerance to moisture and air during operation. Having two valence electrons and high density, zinc metal has three times the volumetric capacity of lithium metal. Among various zinc-based batteries, Zn-air has a theoretical volumetric energy density (6134 Wh/L) that is more than four times of conventional Li-ion batteries (1400 Wh/L). Primary Zn-air batteries have already been the battery of choice for hearing aids, which require extremely high energy density and safety. Finally, zinc is abundant, low-cost, and environmentally benign, rendering them suitable for large scale applications.

Although stable cycling of Zn anodes in mild acidic electrolyte has been demonstrated, an alkaline electrolyte is ideal for zinc-air batteries, because oxygen cathode has minimum overpotential in alkaline electrolyte. However, the performance of Zn anodes in alkaline electrolyte is limited by passivation (**Fig. 1b**), dissolution (**Fig. 1c**) and hydrogen evolution (**Fig. 1d**) problems. Through SEM investigation, critical passivation size was found to be ~ 2 µm. Thus, sub-micron-sized Zn anodes, with feature size smaller than the critical passivation size, won't have passivation problem. As a result, we focus our research on nanoscale. However, Zn dissolution and hydrogen evolution of nanosized anodes will be accelerated because of large electrode-electrolyte surface area. I have designed and fabricated 4 types of zinc anodes to control the interface between the electrode and electrolyte by creating interfaces with ion-sieving and hydrogen evolution suppressive properties. These designs (listed below) have successfully overcome passivation, dissolution, hydrogen evolution problems.

- Graphene oxide-modified zinc anode (Zn@GO, **Fig. 1e**)
- A lasagna-inspired nanoscale zinc anode (ZnO@GO, **Fig. 1f**)
- Sealed ZnO nanorod anode (Zn@TiN_xO_y, **Fig. 1g**)
- A hydrogen-evolution-suppressing anode (**Fig. 1h**)

All these anodes show superior performance compared with unmodified anodes. These anodes can be paired with air cathodes to make high energy Zn-air batteries. The nanoscale design principles here can potentially be applied to overcome intrinsic limitations of other battery materials. Besides, I will also present my recent research progress on operando optical microscopy analysis of Zn anodes. Dendrite growth and the regulation of Zn metal deposition will be visualized.

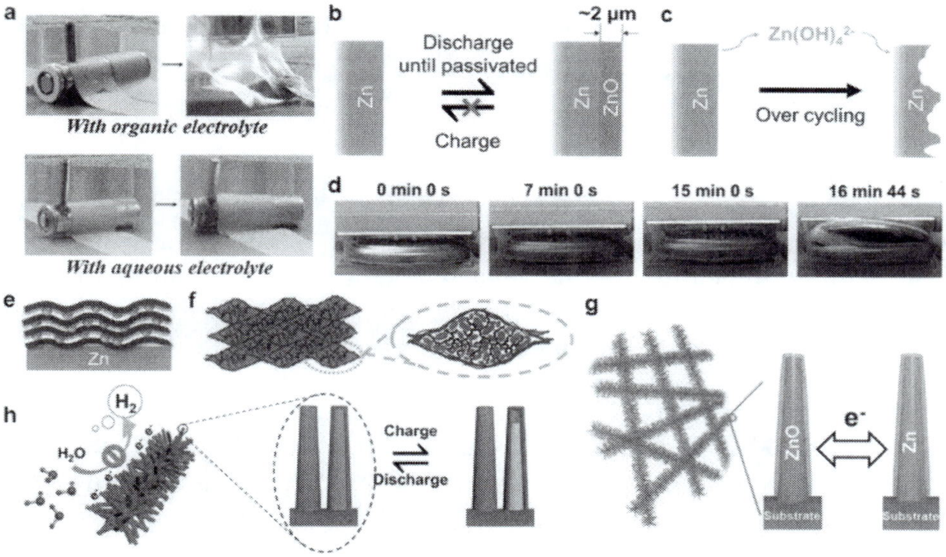

Fig. 1. (a) Battery nail penetration test. (b) Passivation. (c) Dissolution. (d) Hydrogen evolution over time. (e) Zn@GO. (f) Lasagna-inspired nanoscale zinc anode. (g) Sealed ZnO nanorod anode. (h) Hydrogen-evolution-suppressing anode.

608119 The Python Pivot – Teaching Chemical Engineering Computing in the Wake of COVID-19

Abstract:

The COVID-19 pandemic of 2020 caused a major disruption in the education system with the abrupt transition to remote learning. In addition to the quick adaptation of this unfamiliar mode by both students and professors, resources typically used for the course was no longer an option, and course outcomes and deliverables had to be re-imagined.

In this poster presentation, the challenges encountered during this transition and how they were addressed in a junior-level computer simulation course will be discussed. More specifically, the instructor had to rapidly pivot towards teaching Python, because the software used was not available to students remotely. This result in the creation of projects wherein students had to use Python programming to solve three styles of chemical engineering problems that they have been already been exposed to in the curriculum: 1) a multi-step solution, 2) an iterative solution, and 3) an integration. Class time evolved into working periods where groups could interact with their colleagues and instructor within Zoom breakout rooms, and assessments took the form of: 1) a short presentation where groups explained their problem and Python code, 2) a collaborative wiki where each group would describe the Python skills or structures that they needed to solve their problem, and 3) a working Python program that would compute their solution. Student comments regarding the switch to remote learning as well as their experience with this course framework will also be presented.

610679 Collective Realities: A Discussion of Encultured Norms in the Context of Improving Learning and Workplace Justice

Abstract:

Students and faculty have encultured beliefs. These beliefs influence the way we teach and learn, experience our academic institution as a workplace, and create inclusive communities. Through an NSF RED grant, we have been involved in collaborative, focused work on organizational change. In this work, we have observed norms (or encultured beliefs) within the academic community. These subterranean norms guide practices and policies that can hinder improvement of teaching and learning and the cultivation of a just workplace. Our premise is that similar norms exist in other academic communities beyond our own. In this paper, we describe examples of these encultured beliefs and identify how we believe they create barriers to positive transformation of learning spaces and the workplace. We believe such articulation can facilitate discussion that leads to questioning and changing of the underlying norms that consciously or unconsciously guide policy formation and both formal and informal practices.

We took a sociocultural approach, initiated by the authors (who are leaders in several organizational change initiatives) sharing of their professional experiences. We identified common and observed values, and the encultured beliefs that emerged from these values. The authors are deeply integrated members of both engineering culture and that of higher education, and have a shared history over years of reflecting, discussing, and acting on norms propagated within these overlapping spaces, albeit within one particular academic institution. The set of norms that have arisen from this work therefore represent one particular case within one particular community, as seen by one particular group, but we believe similarities exist across academic institutions because engineering culture and that of higher education reach beyond our own lived experience within our institution.

We present examples of norms in engineering (e.g., the belief in meritocracy), norms in the classroom (e.g., authority comes from the instructor and the text), norms of faculty (e.g., mistakes in the classroom show a lack of competence), norms in the academic workplace (e.g., approbation and reward structures that elevate research endeavors above teaching, learning and mentoring contributions) and norms of inclusion and justice (e.g., demographics alone are a good measure of a community's culture as it relates to belonging). We use these identified norms to stimulate discussion of their ramifications.

603987 Engineering Teamwork in Cheme Car: Methods Development

Abstract:

Teamwork is widely recognized as an important skill for engineers in the professional workplace due to the ability for high-performing teams to accomplish more than the sum of their members. The development of teamwork skills in undergraduate engineering programs is included by ABET in their accreditation criteria.[1] Furthermore, recent alumni report that teamwork is among the most important skills in their professional lives.[2] Within engineering education, active learning strategies are being promoted, but the norm in undergraduate engineering classrooms is the passive lecture format. Engineering students are usually evaluated individually and assigned a grade on a curve, in which there is a scarcity of high grades, resulting in student-student competition.[3] This class format runs counter to the development of teamwork skills, leaving a gap in preparation for graduates who enter the workforce in which teams are the standard way to work.[4]

Another way for undergraduate students to gain experience working in teams to solve engineering problems is in undergraduate technical clubs, such as ChemE Car. The ChemE Car competition, run by AIChE, involves university teams building a shoebox-sized model car that is powered and controlled by chemical reactions. One of the objectives of the competition is to provide students with the opportunity to partake in a team-oriented design project.[5] The present research study attempts to assess the development of teamwork skills among members of the ChemE Car team at a large public research institute.

This study is set within the context of professional skill development. We therefore used the Katzenbach and Smith teamwork framework as a basis to observe how the ChemE Car teams work. The fit is based on the professional context of the Katzenbach and Smith model, as opposed to alternate frameworks that are situated within a classroom context with associated instruction and assignments. The features of a team, according to Katzenbach and Smith, are

- A small number of members with complementary skills committed to a common purpose
- Collective work products that multiple team members work on together
- Team members are necessarily working with each other on the same collective work product
- A specific team purpose, which is typically assigned by someone outside the group
- Performance measured by assessing collective work products[6]

This framework is useful for evaluating ChemE Car project teams, because each team is comprised of five to eight people working to improve a power source or control mechanism for the model car, and all the members are working together to produce calibration curves that can be used to control the car at the competition. The task is specified by the conditions of the competition as determined by AIChE, and teams' work products are evaluated by their placement in the competition.

The model provided by Katzenbach and Smith was evolved into a conceptual framework that could be used to evaluate team interactions by Davis and Ulseth (2013),[7] examining four areas of performance: team relationships, production of joint-work products, individual work products, and knowledge assets. This theoretical framework was used to develop the data collection methods to be used in the present study.

The first data collection method we developed is an observation rubric adapted to the specific context of observing a ChemE Car project team working in the laboratory. Observation was chosen as a method because it allows for minimal researcher interference in a self-directed environment, and the researcher can easily collect data on how students interact.[8] The rubric is used to classify the specific actions of teams in the lab as either successful team interactions that promote good teamwork or unsuccessful team interactions that hold the team back. The rubric underwent a validation process, in which it was assessed for relevance to the research question by observing the laboratory sessions of several project teams and coding interactions observed as either within the scope of the rubric or not. It was determined through this validation that team relationships and production of joint-work products could be evaluated through observation, but individual work products and knowledge assets could not be assessed as well.

It was therefore decided that performing interviews with project team members about individual work products and knowledge assets would yield complementary information. Interviews also allow for the researcher to gain insight into subjects' internal thoughts and feelings, which complements observational data related to team relationships.[8] The interview questions were validated by asking ChemE Car members graduating in spring 2020 about the perceived scope and clarity of the questions to ensure the intent of the question is effectively communicated.

The results of this research will be used to write a guide for project managers of ChemE Car teams, with suggestions on how to optimize laboratory time to cultivate the development of teamwork skills among the general members of ChemE Car. In addition, the process of this research will be recorded to provide a process future students can use to investigate the effectiveness of and improve the project managers' guide.

Literature Cited

1. Accreditation Board for Engineering and Technology. (2015). Criteria for Accrediting Engineering Programs. Retrieved from https://www.abet.org/accreditation/accreditation-criteria/criteria-for-accrediting-engineeri ng-programs-2016-2017/

2. Passow, H. J. Which ABET Competencies Do Engineering Graduates Find Most Important in Their Work? Journal of Engineering Education 2012 , 101 (1), 95–118.

3. Smith, K. A. Cooperative Learning: Making "Groupwork" Work. New Directions for Teaching and Learning 1996 , 1996 (67), 71–82.

4. Smith, K. A. Teamwork and Project Management , 4th ed.; McGraw-Hill Education: New York, 2014.

5. American Institute of Chemical Engineers. (2020). 2020 Regional Competition Official Rules . Retrieved from https://www.aiche.org/sites/default/files/media/document/chem-e-car_official_rules_202 0_final.pdf

6. Katzenbach, J. R.; Smith, D. K. Harvard Business Review . 1993.

7. Davis, D.C. & Ulseth, R.R.. (2013). Building student capacity for high performance teamwork. ASEE Annual Conference and Exposition, Conference Proceedings. 8. Mccord, R. E.; Matusovich, H. M. Naturalistic Observations of Metacognition in Engineering: Using Observational Methods to Study Metacognitive Engagement in Engineering. Journal of Engineering Education 2019 , 108 (4), 481–502.

605807 A Video-Based Self-Study Outreach Modeling Project for Future Chemical Engineers

Abstract:

US high school students are falling behind their peers in other developed countries such as Finland and Korea in mathematical performance [1-3]. This presents a need for math-related outreach programs that motivate students to learn more math and introduce them to commonly used scientific tools and formulas during high school. In the last five years, we successfully developed MATLAB-based modeling projects for high school students to build ordinary different equation models for bioreactors like microbial fuel cells [1-3]. In addition to MATLAB Simulink [4], R was also introduced to students as it is a free and popular software widely used in statistics and engineering [5]. In order to deliver these training and research modules to a larger audience, we have summarized all the bioreactor models in a printed book [6]and provided online videos for each example in the book.

Centered around two programs, MATLAB Simulink and R, students are encouraged to learn the basics to ODE models through this self-paced book. The book starts with an introduction to ODE models--as well as an introduction to integrals and derivatives for students who may not have yet been exposed to these subjects. It then establishes the basics of MATLAB for solving ODEs and guides students through different ODE modelling scenarios in increasing complexity, from a running car to a microbial fuel cell [7, 8]. After introducing the scenarios and explaining how to solve them from ODE models in MATLAB, the book then introduces R and describes how to solve the introduced ODE scenarios with R. Recorded videos and codes are provided with step-by-step instructions for each example. Additionally, students complete practice exercises at the end of each example to test their knowledge. At the end, students should be able to solve ODE models with multiple equations and numerous parameters for biochemical processes.

Ten high-school students that completed this book were generally able to complete it within 10 to 30 hours, working independently. They were surveyed anonymously on their experiences with the book. While more high-school students are expected to read the book and finish the anonymous survey in the future, the preliminary survey results showed that: 1) the ideas covered by the book were not too challenging for high-school students; 2) the real life-related models (e.g., microbial fuel cells) made the book more attractive; 3) the book was effective in conveying knowledge on biochemical reactors, Simulink, and R; 4) the Simulink and R instructional components were easy to follow and the simulation results were easy to reproduce; 5) most high-school students would recommend the book to their peers; 6) the book encouraged high-school students to pursue a college major in the STEM fields; 7) and high-school students with little to no understanding of bioreactors or how to use MATLAB Simulink or R codes to solve ODE models can achieve high proficiency after going through the content of the book and the videos. Open-ended feedback from the past readers on the research projects in the book will be further collected and shared. The research modules along with the survey results can provide suggestions for enhancing STEM outreach activities in high schools.

Reference

1. Guo, P., K. Yuan, and Z. Huang. *Develop Web-based Modules to Educate High-School Students in Studying Microbial Fuel Cell Dynamics*. in *2017 ASEE Mid Atlantic Section Spring Conference*. 2017. Baltimore, Maryland.
2. Ekaputra, C. and Z. Huang. *Educating High School Students in Process Simulation and Control*

with a Simulink-Based Controller Design for Microbial Fuel Cells. in *ASEE-SE 2015*. 2015. University of Florida, Gainesville, FL.
3. Geng, J., et al., *Comparison of R and MATLAB Simulink in Educating High School Students with ODE Modeling Skills*. Chemical Engineering Education, 2019. **53**(2): p. 121-129.
4. Su, J.H., J.J. Chen, and D.S. Wu, *Learning feedback controller design of switching converters via MATLAB/SIMULINK*. Ieee Transactions on Education, 2002. **45**(4): p. 307-315.
5. Ihaka, R. and R. Gentleman, *R: a language for data analysis and graphics*. Journal of Computational and Graphical Statistics 1996. **5**(3): p. 299-314.
6. Huang, Z., *Doing Cool Research Projects in Simulating Biochemical Reactors on Computers by a Layperson: MATLAB and R Codes, and Instruction Videos*. 2019, Lafayette hill, PA: IVY STEM Outreach Publishing.
7. Ping, Q.Y., et al., *Mathematical Model of Dynamic Behavior of Microbial Desalination Cells for Simultaneous Wastewater Treatment and Water Desalination*. Environmental Science & Technology, 2014. **48**(21): p. 13010-13019.
8. Ping, Q.Y., et al., *Integrated experimental investigation and mathematical modeling of brackish water desalination and wastewater treatment in microbial desalination cells*. Water Research, 2015. **77**: p. 13-23.

595447 Transcend: Transfer Success Co-Design in Engineering Disciplines

Abstract:

Beginning with the graduating high school class of 2015, the Tennessee Promise program provides "last dollar" scholarships and mentoring programs focused on increasing the number of students at any of the state's 13 community colleges, 27 colleges of applied technology, or other eligible institution offering an associate's degree. In its inaugural class, about 58,000 students (90% of Tennessee's senior class) applied for Tennessee Promise. Thus, the faculty and administration at the University of Tennessee, Knoxville (UTK) anticipate that transfer students are likely an increasingly important fraction of the student body. In the Tickle College of Engineering (TCE), transfer student enrollment has increased at an average rate of 10% per year over the last several years. Transfer students face a unique set of challenges, which differ from those who begin their first year at a university as freshmen. The challenges are both academic--adapting to the reduction in individual attention with university coursework--and social--lacking a well-defined peer-cohort with whom the adjustment to university life can be shared. In engineering, transfer students also statistically represent a different demographic population than the student body of entering freshmen (the fraction of first-generation college students is double (32% compared to 15%) among transfer students, which can potentially bring cultural challenges as well. These challenges manifest in an unfortunate fact: the fraction of engineering transfer students who do not graduate within 5 years is nearly double (29% to 15%) that of traditional engineering students who have an analogous two years of college behind them. Finally, existing programs for financial aid are disproportionately distributed to students who enter the university system as freshmen relative to transfer students; the unmet annual financial need of engineering transfer students is more than double that of traditional engineering students at UTK (>$6,000 to ~$3,000).

The TranSCEnD: (Transfer Success Co-Design in Engineering Disciplines) program is in its third year at the University of Tennessee. It is funded by an NSF S-STEM grant with an objective to increase the retention of engineering transfer students to a level comparable to engineering students, who entered the TCE as freshmen. To achieve this task, a multidisciplinary leadership team has been assembled, which includes UTK tenured professors, lecturers, staff, and administration from TCE, staff from the UTK admissions office, shared faculty from Pellissippi State Community College (PSCC), the single largest source of engineering transfer students (nearly half) to the TCE, and an external assessor. The principle of codesign, namely the explicit inclusion of input from all constituents, has been invoked in the creation of the TranSCEnD program. The program spans a five-year process--two years at the community college, a summer bridge program, and three years at UTK. The individual elements of the program as well as the synergistic integration of elements have been chosen to balance two influences: (1) a theoretical framework of best practices and (2) a practical acknowledgment of demonstrated success at UTK. The program offers financial, academic and social support for transfer students in the TCE. Program elements include a summer transition research experience, a student success skills seminar and intra-cohort and inter-cohort peer interactions. Reflections on the program success and challenges to date will be shared.

597986 Quality Improvement in Graduate Chemical Engineering Kinetics

Abstract:

Often graduate students in chemical engineering come from diverse B.S. degrees beyond just chemical engineering. These majors include the sciences (e.g., chemistry) and other engineering disciplines (e.g., mechanical, materials, environmental, and biomedical). Often the knowledge bases of these fields greatly differ. Chemists often lack the math preparation, whereas mechanical engineers lack the chemistry and materials engineers are not used to working with gases. This quality improvement project focuses on leveraging pathways to enhance student pre-requisite knowledge. Student-centered online activities were created and developed to improve student success and engagement in their STEM courses. In collaboration with Learning Designers from the Innovative Education's Digital Learning team, the instructor developed online activities rooted in best practices and leveraged impactful instructional technologies. Using layered online content and integrated quizzes with feedback, the prerequisite knowledge (or need activities to refresh after being away from fundamental chemical engineering concepts). As a result, students' performances were on par with those of previous years without devoting valuable class time on those prerequisite concepts; therefore, taking time away from the intended course content. Additional qualitative and quantitative comparisons will be presented, as well as plans to expand this needs assessment project into the next academic year.

610170 Colormephd: Communicating Research to a Broad Audience through Coloring Pages

Abstract:

Communicating science to the general public is a challenge, and chemical engineers at various stages in their careers often struggle to communicate the results of their research papers to a non-scientific audience without compromising the integrity and rigor of their research. ColorMePhD is a free online coloring book that describes current PhD-level research in chemical engineering and related fields, using a combination of visual elements and descriptions written in accessible language. The goals of this initiative are 1) to provide a free resource for teachers, students, members of the public to engage in research, 2) to provide diverse science role models for aspiring scientists, and 3) to provide undergraduate student researchers, graduate students, postdoctoral researchers, and professors with an opportunity to develop their skills in communicating their research to a broad audience. This poster details the process for the development of the coloring pages, responses from scientists who participated in the program as featured scientists, and analysis of social media engagement with the content. The response from participants of a workshop held at the University of California, Berkeley, entitled, "Communicating Your Research to a Broad Audience: Methods Inspired by STEM Outreach" will also be presented, along with examples of visual metaphors and analogies used to communicate complex research topics in the coloring pages, as well as insights on the development of inclusive analogies, metaphors, and language.

603288 *Enhancing the Department of Energy's Industrial Assessment Center Experience for Undergraduate Students through Real-World Problem Solving, Research, and Publication*

Abstract:

There are 31 universities nationwide that operate Department of Energy Industrial Assessment Centers (IACs). The centers develop student-led teams that are trained to analyze the energy consumption of manufacturing processes [1]. The teams put those energy management skills to work when they visit small and medium-sized facilities located in the IACs' region [2]. These facilities primarily focus on their manufactured products, and generally do not have dedicated energy management personnel on site. The IAC students fill the energy management role and develop a host of ideas to help the facilities minimize their utility costs and increase process efficiency. This is a powerful framework for students to gain industrial engineering experience and to provide real-world value to the manufacturing sector.

The Intermountain Industrial Assessment Center located at the University of Utah is organized with two faculty members at the head, four graduate students, and ten undergraduate students. Faculty members facilitate the connection with manufacturing facilities and quality control the final energy management report. Graduate students and experienced undergraduate students serve as energy assessment leads. Newer undergraduate students learn critical energy management skills with targeted engineering tasks and calculations performed for specific recommendations on each assessment. Common recommendations include, upgrading to LED lighting, installing variable frequency drives, and performing heat recovery.

Sometimes IAC site visits yield cutting-edge energy saving or process enhancing recommendations that can provide academic value beyond the scope of the report. When these ideas are coupled with driven undergraduate students who desire to elevate the level of their engineering analysis, an opportunity for academic research arises. Over two years, three undergraduate students have been led by a graduate student and faculty member through the process of transforming novel IAC ideas to solve real-world problems into the undergraduate students' own research [3].

As with any educational experience, finding a balance between support and micro-management of undergraduate students during the industrial research process was critical. Fortunately, working outside of the classroom allowed for highly tailored mentorship. The general recipe was a high frequency of meetings early on to get student investment into the project and to demonstrate support. This was paired with the vivid real-world problem that was encountered on the recent facility site visit. Through trial and error, each research project then found a natural frequency for meetings as the project became more fleshed out. The content of the meetings was customized to give the student the tools or direction they lacked to model or analyze the problem without explicitly solving it. They were then assigned manageable action items to perform prior to the next meeting. More specific action items led to a higher probability of success. These accomplished tasks, strung together, developed into research papers. The open-ended nature of research can be daunting. With regular personal support, limited technical support, and big picture guidance, three undergraduate students were able to transform interesting real-world industrial solutions into publications [4-6]. For each student this was their first publication. It was important to ensure success, not guaranteed with undergraduate research, and help them get their feet wet with this novel, exciting experience [7].

[1] Dalzell, Nicole M., Gale A. Boyd, and Jerome P. Reiter. "Creating linked datasets for SME energy-assessment evidence-building: Results from the US Industrial Assessment Center Program." Energy Policy 111 (2017): 95-101.

[2] Abbas, Ahmad I., et al. "A Comparative Study of Industrial Energy Assessments for Small and Medium-Sized Industrial Facilities." ASME 2018 12th International Conference on Energy Sustainability collocated with the ASME 2018 Power Conference and the ASME 2018 Nuclear Forum. American Society of Mechanical Engineers Digital Collection, 2018.

[3] Dolan, Erin, and Deborah Johnson. "Toward a holistic view of undergraduate research experiences: An exploratory study of impact on graduate/postdoctoral mentors." Journal of Science Education and Technology 18.6 (2009): 487.

[4] Westberg, B., et al. "Proactive Automation of a Batch Manufacturer in a Smart Grid Environment." (2018): 110-131.

[5] Henning, Moriah, Derek Machalek, and Kody M. Powell. "Integrating a microturbine into a discrete manufacturing process with combined heat and power using smart scheduling and automation." Computer Aided Chemical Engineering. Vol. 47. Elsevier, 2019. 293-298.

[6] Brimley, Paige, Derek Machalek, and Kody M. Powell. "Smart Scheduling of a Batch Manufacturer's Operations by Utilization of a Genetic Algorithm to Minimize Electrical Demand." (2019): 53-67.

[7] Alsuhaibani, Marya, et al. "Research education in an undergraduate curriculum: Students perspective." International journal of health sciences 13.2 (2019): 30.

600643 Plate Spinning: A Beginners Guide to Excelling in Academia As an Engineer/Scientist

Abstract:

A vast majority of faculty members (also called academic staff) in western Universities arrive at their first academic position following the established progression from undergraduate to postgraduate and some postdoctoral study. This route prepares the new faculty colleagues well for executing research and elements of teaching but falls desperately short of the preparation required for modern academia. Often newly appointed faculty will face a steep learning curve as they adopt the many functions required to succeed in their new roll.

The challenges placed upon newly appointed faculty are tremendous. The struggle for credibility in research whilst balancing a teaching and administration role, in addition to managing an often young family, is a significant challenge. Meeting early career challenges to become a research leader is perhaps the most difficult stage of the modern academic career. This presentation provides a gentle steer towards taking a strategic approach by adopting a new view: thinking as a CEO of a small enterprise while defining a vision and constructing a pathway to achieve that vision. We will share the advice which benefited the authors and it forms the basis of discussion. We will introduce aspects that the authors found useful in their journey, such as mapping research niche, differentiating between aims and tools, building resilience, the "*research transfer function*", networking and marketing. This frame of reference and a quick-start guide which is intended to facilitate new appointees in Science and Engineering to make sense and success of their first 5 years following appointment.

About the Authors

Siddharth and Adam, while still relatively early career academics, have enjoyed some success in their teaching, administration and research. However, their progression has been fraught with misadventures, failings, good/bad luck and lots of hard work. They have compiled this information to help new faculty in science and engineering to obtain the correct balance of the aforementioned. Adam is a Rolls-Royce/Royal Academy of Engineering Research Chair and the Director of Admissions for the Faculty of Engineering. He is also a Director of and Scientific Advisor to Texture Jet Ltd which is a University spin out company. Siddharth is Professor of Sustainable Chemical and Materials Engineering and EPSRC Fellow in Manufacturing. He chairs the Faculty Search Committee and recently as a chair of the Curriculum Committee, lead the transformation of chemical engineering education at Sheffield.

612375 Leading and Managing Industry-Academia Partnerships and Collaborations Panel Discussion

Abstract:

Outline:

- Discussion on benefits resulting from and concerns hindering industry-academia collaborations
- New methods and long term changes needed to lead and manage industry- academia partnerships and collaborations
- Benefits to diverse sectors including R&D, manufacturing, economics, strategic planning etc. and increasing participation across all.
- Definition of research contracts, communication channels, international collaborations, project timelines and deliverables

Invited Panelists:

Barclay Satterfield: External Innovation Leader at Eastman Chemical Company who has experience in open innovation and industry/university research partnerships, polymer property research and environmental life cycle analysis. She has been involved in a $10M 6-yr research partnership between Eastman Chemicals and NC State University.

Kenneth Rueter: President and CEO - UCOR LLC, a past Management Award recipient has over forty years of experience. He has been involved in a highly successful collaboration with the University of Tennessee Nuclear Engineering Department. This collaboration has established a Nuclear Decommissioning minor at the University of Tennessee enabling students to meet the high demand for Nuclear Legacy Cleanup around the World.

Dr. Kiarash Vakhshouri: Young professional working with Google with a solid expertise in structural characterization of semiconductor materials, thin film coating and printing, rheology, lithography and designing and operating laboratory and pilot plant chemical processes. He has several well-cited publications and four US patents to his credit and has been engaged in a number of industry-academia collaborative research projects right since graduate school.

Dr. Phil Westmoreland: Has a PhD from MIT and has been a Professor of Chemical and Biomolecular Engineering with the North Carolina State University. He has over 35 years of academic research experience and has held a variety of Directorial positions during his illustrious career. He has also served as the AIChE President in 2013. His research interests include flame chemistry, theoretical and computational chemistry and polymer decomposition kinetics and thermochemistry. He has been instrumental to a number of industry-academia-government collaborative projects.

Dr. Quinta Warren:CEO at Energy Research Consulting, PE, PMP has experience in industry, academia as well as Government. She was working with ConocoPhillips, has been a director of professional education and also worked with the US DOE. She brings in a unique entrepreneurial skill-set into synergizing these areas.

Dr. Sarma Pisupati: Professor of Energy and Mineral Engineering/Chemical Engineering and Director of Center for Critical Minerals at Penn State. He has more than 35 years experience in academia and has prior industrial experience as well. He specializes in computational fluid dynamic modeling of combustors, coal/biomass gasification, and biomass based materials for emission reduction. He has driven several successful collaborative research projects and novel initiatives with the US DOE as well as industrial partners.

Samuel Tam:(Have reached out only this week and waiting to hear back) President and CEO of CAC Solutions LLC and has been with the US DOE for 9 years where he managed total annual R&D spending was over U.S. $100 MM. He also has experience as a China Chief representative while working with the US DOE. Over his forty years of R&D, engineering, business development, pilot plant design and operation, and commercialization of technologies related to climate change in US DOE and five other major US corporations, Samuel has gained immense experience on fostering collaborations.

604999 Redesigning a First Year Introduction to Chemical Engineering Course to Integrate a Laboratory Component

Abstract:

A one-semester introduction to chemical engineering course was redesigned to add a laboratory component. The goals of the redesign were to provide the following for the students: an opportunity in which they could feel an intradepartmental connection working in small groups, basic exposure to the challenging concepts which they will face throughout their academic careers, and hands-on experience to both engage them and increase their retention of the material.

The structure of the course was such that up to 144 students could be enrolled. A rotation schedule was implemented so that each student experienced four, two-hour labs during the semester. The laboratory component of the course focused on the production of biodiesel, with each experiment highlighting a different part of the process. In doing so, the students were able to be exposed to many engineering concepts and clearly understand the integration of each concept/unit into the bigger picture of an engineering process. During the first week, students produced biodiesel from vegetable oil using one-liter batch reactors. The products were separated by gravity and then ion exchange was used to purify the biodiesel in the second lab. During the third lab, the students used distillation to recover excess methanol from the glycerin produced in the biodiesel reaction and then compared the results to an ASPEN simulation. Finally, during the fourth week, the students produced soap using glycerin from biodiesel production, tested the biodiesel's cold weather properties, created a 20% blend of biodiesel and diesel, and then used the fuel in a diesel generator to make microwave popcorn. During each lab, the students also interacted with larger-scale equipment for each process unit being studied.

In addition to the laboratory component, students had a 50-minute class each week in which they engaged with a combination of technical and departmental content, including a discussion on safety and ethics. Faculty members and seniors in the department held roundtable discussions, answering questions about research, internships, and potential career paths. The classroom was specifically designed for active learning with students sitting in small tables to facilitate discussions and group work. Problem solving and discussions were guided by the instructor, graduate teaching assistants, and two undergraduate teaching fellows.

The final project for the course was designed to integrate the students' laboratory experience and classroom knowledge. Students were required to design a poster on a product or process of their choice and then present that poster during a poster session. Some of the requirements included presenting a process flow diagram and a discussion of how the class concepts were reflected in the process they selected. Peer evaluation was utilized along with evaluations from the instructor, graduate teaching assistants, and undergraduate teaching fellows. The poster session was held in the atrium of the engineering building and was attended by faculty, students, and staff in the School of Engineering.

Retention analysis of first-year students to the university from the first to the second semester shows an increase of approximately 6% with the redesigned course. Additional data on retention to the major and the evaluations from the students will also be presented.

611631 It's All By Design: Incorporating Design Early in the Chemical Engineering Curriculum

Abstract:

In the early years of their Chemical Engineering education, students can feel overwhelmed by the difficult topics being introduced. Simultaneously, a general lack of understanding of how some early courses tie in with their ultimate career goals can discourage students from continuing with the major. U.C. San Diego has incorporated a design-based group project early in its Chemical Engineering program to facilitate learning and promote interest in the curriculum as a whole. The students are introduced to a detailed chemical process and analyze the system while making various design considerations. In this talk, an example of the project is highlighted and is discussed in terms of promoting groupwork and improving student understanding. Likewise, how these projects satisfy several ABET Student Outcomes will also be addressed.

611944 Integrating Entrepreneurship Mindset through Makerspace-Based Design Project into a Second-Year Engineering Curriculum

Abstract:

In today's global chemical engineering industry, softer skills like teamwork experience, presentation skills, awareness of social and environmental responsibilities, and understanding of global impacts are now recognized as crucial. Practicing chemical engineers are expected not only to be technically knowledgeable but also skilled in communication and leadership. Along with soft skills, entrepreneurship mindset (EM) with the characteristics of curiosity, connections, and creating value has become one of the skillsets that employees are screening for. However, the typical chemical engineering undergraduate core curriculum has not adapted to satisfy all these needs in the chemical industry. Integrating opportunities into the curriculum for students to develop these skills is essential but challenging. To overcome this challenge, many universities have been opening academic Makerspaces on college campuses.

Makerspace equipped with digital design and manufacturing tools, along with traditional machine shop equipment, allows the student to develop engineering design skillsets. The primary purposes of Makerspace facilities are to foster students' creativity, to promote hands-on prototyping and designing, and to encourage cooperation; as a result, to create a community. However, many freshman and sophomore-level students are not aware of the opportunities in the Makerspace. Thus, it is essential to introduce early-year students to Makerspace and encourage them to utilize the space. To achieve this, "designing a calorimeter" project was integrated into the "Introduction to Thermodynamics" course. This course is a sophomore-level class and is the first course in the chemical engineering sequence. The assigned project included background research phase, 3-D modeling phase, prototyping phase, and testing phase. Throughout the semester, students need to work with their team and delivered project milestones, created Adobe Spark page as their final artifact, and engaged in a competition at the end of the semester. This project improves student entrepreneurship and creativity skills. It helps them to understand design phases in the early year of their degree as well as develops their teamwork and communication skills.

In this presentation, the details of the project will be described, and the learning outcomes will be evaluated. The author believes similar assignments can be easily implemented into other courses resulting in sustaining Makerspace and developing students' skillsets.

607168 Using the Renaissance Foundry Model for First-Year Student Engagement through High Impact Practices (HIPs)

Abstract:

The purpose of this contribution is to provide the strategies for integrating High Impact Practices (HIPs), with the Renaissance Foundry Model (herein, the Foundry) – an innovation-driven pedagogical platform (Arce et al., 2015) – as implemented in a freshman-level ethics course in the Department of Chemical Engineering at Tennessee Technological University (TN Tech). Specifically, the student creation of e-Portfolios (an HIP) was leveraged in a manner that encouraged students to assess engineering designs through various lenses (e.g., global, economic, societal, environmental and cultural) through an iterative process linked to the Foundry. The Foundry, which provided the platform for which students engaged in the e-Portfolio throughout the semester, is an iterative pedagogical learning platform which comprises two major learning paradigms: knowledge acquisition and knowledge transfer (Arce et al., 2015). Towards the creation of a prototype of innovative technology, students must engage in the six elements of the Foundry while iteratively progressing through both the knowledge acquisition paradigm and the knowledge transfer paradigm. In utilizing the Foundry towards the development of an e-Portfolio, student teams focused on an engineering design while leveraging various lenses for analysis that was guided by the Social Construction of Technology (SCOT) framework that includes four major criteria: interpretive flexibility, wider context, relevant social groups, and closure and stabilization (Pinch & Bjiker, 1984). In the implementation of this course redesign, there were three major takeaways in terms of implementing an HIP in this context, specifically as it concerns the e-Portfolio project. For the purpose of this contribution, these three takeaways include: how team dynamics influenced engagement, technological implications in pedagogy, reflections on the learning curve. Key observations from these items included the need for contracts, negotiations, and conflict mitigation in addressing elements of team dynamics, the necessity of pedagogical support during the creation of the e-Portfolios for students to gain initial training in using the Google Sites platform for their assignments, and finally leveraging the iterative nature of the Foundry implementation as related to these assignments for web design allowed students and instructors to navigate the learning curve associated with the integration of new technology. The three takeaways presented herein have provided opportunity for reflection and guidance for future iterations of this course.

592674 Some Reflections upon Project-Based Learning in a Traditional Fluid Mechanics Course at Njit

Abstract:

Engineering education has strong emphasis on providing students with analytical, problem-solving skills [1]. Most jobs requiring these skills can now be done better and/or cheaper by either computers or skilled workers in developing countries [2-4]. Hence, engineers with unique, soft skills such as creativity, strong interpersonal skills, holistic-multidisciplinary thinking, cultural awareness, effective communication and leadership skills, etc. will continue to find jobs in the new global economy [2-3]. While general education requirements in universities address development of literacies that help building such skills, discipline-specific approaches in engineering education are also warranted. This paper presents some reflections from the incorporation of project-based learning in the traditional fluid mechanics course CHE260 "Fluid Flow" in the sophomore year of undergraduate chemical engineering program at NJIT. First, based on a walkthrough of our program, the need for incorporating "soft skill" development in our sophomore chemical engineering courses is indicated. Then, using the Fluid Mechanics course as a pilot course, the feasibility of project-based learning opportunities in sophomore courses is demonstrated. In Spring 2019, CHE 260 students (n=44) were asked to write a six-page report that describes, analyzes, and discusses a technology, process, or product, which had a major impact on society, environment, health, a whole industry sector at a global scale either positively or negatively or both. Students were also asked to speculate on ways that negative results might have been avoided. Moreover, students were also asked to explain how the impact specifically relates to the concepts, equations, and approaches of fluid statics and fluid flow or rheology of the fluids, suspensions, etc. involved. Based on the above aspects, rubric attributes were developed, and student learning outcomes were assessed. The assessment and evaluation of the student learning outcomes suggest that project-based learning can be incorporated even in a hard-core, foundation course like Fluid Mechanics with the goal of enhancing students' soft skills such as effective writing, recognition of the professional responsibilities and of the broader impact of engineering solutions on society, environment, and economy, and the use of new tools for acquiring new knowledge and its application to a particular project. This talk will also present what worked and what did not work as well as some recommendations for continuous improvement of the project-based learning and curricular and institutional approaches that can further support project-based learning in the chemical engineering program.

References

[1] R.M. Felder, A whole new mind for a flat world, Chemical Engineering Education, 40(2), 2006, 96-97.

[2] T.A. Friedman, The World is Flat, New York: Farrar, Straus, & Giroux, 2005.

[3] D.H. Pink, A Whole New Mind, New York, Riverhead Books, 2005.

[4] 2005 NAE Report, Offshoring and the Future of U.S. Engineering: An Overview.

608048 Comparison of Problem-Solving Skills on Textbook and Youtube Problems across Two Cohorts

Abstract:

Problem solving is a vital skill of engineers. Previous studies have introduced a novel teaching strategy that replaces homework problems from textbook problems with student-written problems inspired from watching YouTube videos. Replacing textbook problems with YouTube problems, this research focuses on examining the rigor of YouTube problems as well as students' problem-solving skills on textbook and YouTube problems. A quasi-experimental, treatment/control group design was employed and data was collected and evaluated using multiple measurement instruments. Tailoring focus on a topic involving material and energy balance for reacting systems, rigor of problems was examined using the NASA Task Load Index. Also, problem solving was assessed using a previously-developed rubric called PROCESS: Problem definition, Representing the problem, Organizing the information, Calculations, Evaluating the solution, Solution communication, and Self-assessment. PROCESS was modified to independently measure completeness and accuracy of student responses, as well as identify errors committed in material and energy balances. New results comparing two cohorts solving YouTube and Textbook problems will be explored.

611045 Exploration of a Non-Traditional Assessment Method

Abstract:

The COVID-19 pandemic in Spring 2020 catapulted the nation into social distancing to slow the spread of the virus. In response, institutions of higher education pivoted to remote learning. While many brick-and-mortar classes were conducted as flipped classrooms or in other formats that allowed some remote learning, student assessments like exams could be administered face-to-face. The pivot to fully remote learning motivated the exploration of alternative assessment methods that accurately measure student learning outcomes, encourage creativity, eliminate the need for proctoring and don't require face-to-face administration. In response to this need, a non-traditional approach to assessment in which students were expected to write an exam along with the solutions in place of a traditional exam was explored. The course is Fluid Mechanics, a sophomore level course in the chemical engineering curriculum. The instructor provided the students a detailed rubric and a practice exam. In earlier exam administrations, the practice exam and solutions gave the students the freedom to assess their own readiness for the actual exam. In the non-traditional exam, the practice exam served as a model exam, along with the rubric, to assist students in preparing a quality submission. In the referenced course, two of three regular exams were administered in a more traditional format before the pivot to remote learning. Thus, the previous two exams provide a baseline to compare student performance. The results of the study will be discussed.

609535 Learning Assistants: Learning about Learning

Abstract:

There has been recent interest in the incorporation of undergraduate learning assistants (LAs) to support, catalyze and enhance active learning in STEM classes. LAs are undergraduate students who have typically completed the particular course and return to assist with its instruction. Practicing LAs increase interactive engagement of the students in active learning classes by providing near-peer help. The processes of facilitating student learning are also construed as a learning experience of LAs themselves.

The LA Program utilizes the three core elements suggested by the *Learning Assistant Alliance* (https://www.learningassistantalliance.org/). First, in the LA Pedagogy Seminar, LAs receive pedagogical development in in a formal class with their peers, generally in their first term as an LA. Second, LAs meet weekly with the instructor and the graduate teaching assistants as a member of the instructional team to prepare for active learning in class that week. While LAs elsewhere are often used in large lecture sections, in the context of the unit studied, the LAs facilitated learning in smaller studio or laboratory sessions. Third, LAs facilitate active learning in the class in which they are assigned.

In this paper, we briefly describe the LA Program at Oregon State University, focusing on the weekly LA Pedagogy Seminar. We then discuss a reflection tool where students are directed to respond to a weekly prompt and an instructor interface to select prompts and to provide feedback. The tool was piloted and developed iteratively over two years in the LA pedagogy seminar and is available for faculty to use in their courses through the Concept Warehouse (cw.edudiv.org). Finally, wea present research on the changes in the ways undergraduate LAs conceive instructional practice.

At the start of the term, LAs responded to the prompt, "*Think about teaching and learning in the environment that you are working in as an LA (e.g. studio , lab, ...). Identify two or three strong beliefs that you have about your role to support learning in that environment.*" Thematic coding is used to characterize the aspects that the LAs identify as salient to teaching, the ways their thinking has shifted, and what concepts from the LA pedagogy seminar are being incorporated into their thinking. We ask the following research questions:

1. What are the common elements of the beliefs that LAs bring to instruction? How do LAs differ in their expressions around these beliefs?
2. How have the expressed beliefs shifted over the first term of the LA experience? What concepts from the LA pedagogy course are incorporated into the second reflection?

While research on LAs has addressed gains in their technical knowledge as a result of instruction, this research contributes to understanding other ways LAs benefit from their experience.

605716 Correlating Student Success and Attempts on Auto-Graded, Randomized Homework Questions for Material and Energy Balances

Abstract:

While an interactive textbook for material and energy balances, published by zyBooks - a Wiley brand, has shown excellent student engagement with median reading rates as high as 99% over the past 5 years, using interactive platforms for auto-graded randomized homework has been available for some chemical engineering courses for about a decade. Three types of auto-graded problems will be explored, namely multiple choice, single numeric answer, and multiple numeric answers. With over 500 questions across 9 chapters, more than 80,000 attempts were logged for ~100 students in recent years. No penalty is incurred for an incorrect response, so student can persist until correct and re-use problems for additional practice later in the course. The auto-graded problems are due weekly and account for 5% of the student's semester grade. Student success in the course, such as final course letter grades, will be correlated with the fraction of auto-graded homework questions answered correctly before the due date. In addition, question order and question type will be examined with respect to attempts before correct. Overall, big data is generated with interactive online homework tools that can help understand student's engagement, persistence, and success in chemical engineering courses.

599206 Creating a Confident and Curious Cohort: The Effect of Video-Led Instruction on First-Year Chemical Engineering Students Practical Learning

Abstract:

INTRODUCTION

Widely adopted in education, the experiential learning theory states that "learning is a holistic process of adaptation to the world... thinking, feeling, perceiving, and behaving" (Kolb and Kolb, 2009). This serves as the theoretical foundation for the laboratory learning modules in chemical engineering education (Chen et al., 2016b; Cline et al., 1997), where students gain first-hand experience with data collection and analysis through specialised equipment and tools.

The ubiquitous use of computers and the Internet has enabled educators to enhance experiential learning with e-learning materials such as digital notes and instructional videos. By sharing such resources online, educators in the USA have started a revolution with the concept of massive open online course (MOOC) (Hew and Cheung, 2014; Hyman, 2012). Literature implies that hands-on laboratory modules are still the most effective in delivering experiential learning.

This study reports a strategy that has been implemented in the Undergraduate (UG) Teaching Labs of the Chemical Engineering Department at Imperial College London, UK, for employing videos as an instruction tool delivery of a hands-on laboratory module while enhancing independent learning. At the core of this strategy are the instructional videos that students have to watch before the experiments. These videos reduce the training time that teaching assistants have to spend on each group of students and hence enhance the quality of intellectual engagement and helping the students to become much more independent learners. The paper aims to share good practices that are based on observations and survey results, so that interested educators can readily adopt our strategy according to their needs.

Delivery of Instructional Videos

From the academic year (AY) 2016/2017 the teaching of 1st year UG labs (Foundation Labs) was supplemented using instruction videos. The Foundation Lab consists of 7 experiments conducted by the students in groups of 2 or 3 on a rotational basis. Prior to AY 2016/17, the groups were provided with a written handout for each experiment. Each experiment was also assigned a graduate teaching assistant (GTA), who is tasked with guiding the students through the experimental process. From AY 2016/17, the use of instructional videos sits alongside the written handouts and the GTAs to give the students further guidance before and during carrying out the experiments.

Initially students were not provided with access to instructional videos, only the handouts, and GTA supervision. This was to provide a control, against which student engagement and learning outcomes were compared to when the videos were in use. After 3 cycles of experiments videos were made available to watch any time for the remaining duration of the course.

RESULTS & DISCUSSION

Student Perception

Anonymous polling of the students via Mentimeter found the students' viewed the use of video on-demand positively (See **Figure 1**).

While responses to the videos were largely positive, some free text comments revealed how video-led instruction could be adapted to improve the implementation.

Some students did find the instructional videos useful, but only for pre-laboratory instruction, rather than a replacement for any current systems. This fits well with how the videos were presented to the students, as students were instructed to watch videos before laboratory sessions. Another respondent who did not think the instructional videos improved their learning experience said:

"The videos should match the handouts because sometimes they contradict each other"

However several respondent indicated that the videos should contain more information and expand beyond the scope of the handouts in some cases:

"Perhaps give more hints and tips that aren't on the handouts."

The students revealed in a focus group discussion that the videos helped them gain confidence before attending the labs, arriving better prepared.

Student responses indicate that the videos act as supplementary material to the handouts, adding to the student experience. Students stated that "watching the videos saved time, and some students felt more efficient with their lab time, however the importance of having a GTA in the lab available was helpful".

GTA Perception

GTAs noticed a marked difference in the output of the students once videos were released, noting that risk assessments (the first assessed work for each experiment) improved after the introduction of video instructions and were better than student cohorts from previous years. GTAs noted "a change in students' behaviour, they started to run the labs more independently after watching the videos." This is in keeping with the student comments that videos made the student feel more independent. GTAs suggested "there should be a clear differentiation between demonstrational/operational and problem-solving videos." GTAs suggested that problem-solving videos could be produced to address common operational issues students have, "freeing GTAs' time, making them available for further discussion of theory".

Course Co-ordinator Perception

Interviews were carried out separately for the Principle Teaching Fellow and Lab Technician to understand their perception of the Impact operational Videos have on teaching.

Based on their perceptions of the student behaviour once the videos had been implemented (after the 3rd experiment) highlighted that students came to the labs "more prepared to operate the equipment" and as students knew more about the operation of the equipment, it made GTAs more free to act in "an advisory" role.

"They appeared a lot more confident. You feel that they are more prepared, as they've seen the equipment on film and straight away you can see themselves familiarising themselves with the equipment, and comparing it to the video." - Andrew Macey, Lab Technician

"Rather than asking them [GTAs] operational questions they were asking them more intellectually challenging questions" - Umang Shah, Principal Teaching Fellow

By allowing students to focus less on operational issues, the course co-ordinators could also improve their own teaching.

"Students were asking [me] a lot of engaging questions, above and beyond the project scope and that pushed me to think about how we can make it more open ended, because students were starting to take the interest, they were becoming more engaged and more independent" - Umang Shah

The videos also contributed to making organisation of the labs "much more straightforward and simpler to co-ordinate". Students came to labs better prepared; more students were able to "finish [their experiments] with 3 hours". Videos also helped standardise the teaching of the operational aspect of the equipment in the labs.

CONCLUSIONS

This study highlights a paradigm change in experiential learning. Video-led instruction in a practical learning module enabled students to take charge of their own learning, as well as freeing up valuable education time. This allowed for more intellectually stimulating discussions regarding the projects, their contexts, and wider applicability of the work. Students engaged well with shorter videos, and multiple short length videos covering specific aspect of the experiment were found to be more effective. This study demonstrates the potential for employing videos to transform the role of lab demonstrators to facilitators and ultimately mentors, thereby creating a cohesive enabling environment for independent learning.

REFERENCES

Chen, W., Shah, U., Brechtelsbauer, C., 2016b. The discovery laboratory A student-centred experiential learning practical: Part I Overview. Educ. Chem. Eng. 17, 44-53.

Cline, M., Powers, G.J., Ep, I., 1997. Problem based learning via open ended projects in Carnegie Mellon University's Chemical Engineering undergraduate laboratory. Front. Educ. 1997 - 27th Annu. Conf. Proceedings, Bols I - Iii 350–354.

Hew, K.F., Cheung, W.S., 2014. Students' and instructors' use of massive open online courses (MOOCs): Motivations and challenges. Educ. Res. Rev. 12, 45-58.

Hyman, P., 2012. In the Year of Disruptive Education. Commun. ACM 55, 20–22.

Kolb, A.Y., Kolb, D.A., 2009. Experiential learning theory: A dynamic, holistic approach to management learning, education and development. SAGE Handb. Manag. Learn. Educ. Dev. 42–68.

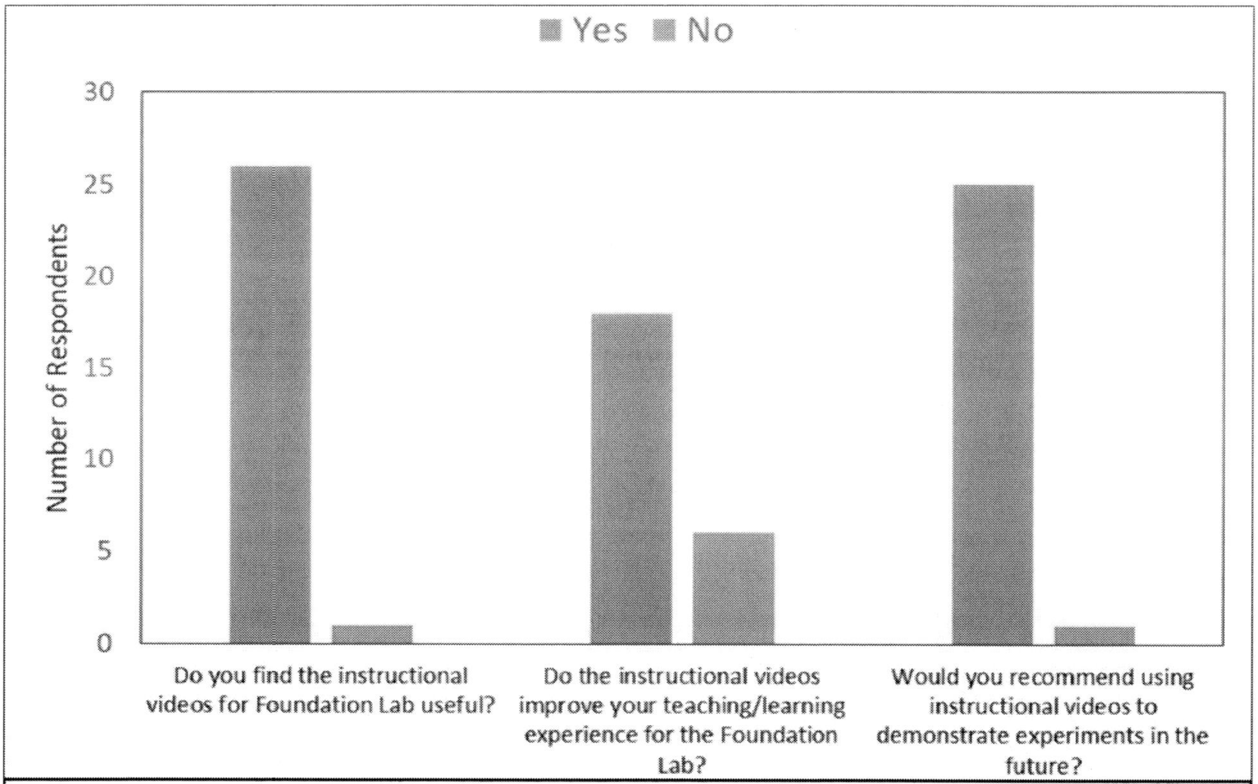

Figure 1: Mentimeter polling of Students' opinions on the implementation of video-led instruction in the Foundation Labs.

600335 A Software for Learning Quantitative Critical Thinking By Chemical Engineering Students

Abstract:

Critical thinking is an important component of many medium-complexity and high-complexity jobs in various industries. It is also an important life-skill that is valuable to the individual as well as the workforce at large. Despite the importance and interests ascribed to this life-skill by governments, employers and educators, the teaching of critical thinking is seldom infused explicitly into the curriculum of an undergraduate programme. Based on the result of a covert pedagogical experiment conducted on second-year Chemical Engineering students in the initial phase of the current study, an important gap in engineering education in helping students develop higher order thinking skills was apparent. Engineering graduates may not be adequately equipped with the ability to apply critical thinking within the engineering domain and towards quantitative evaluation of engineering systems, referred to as quantitative critical thinking. The approach of repeated practice was applied in this study to develop an intervention for enhancing quantitative critical thinking skills by Chemical Engineering students. The intervention is in the form of a software, referred to as a Quantitative Critical Thinking (QCT) software, developed using the Matlab programming language. This software is programmed to generate detailed calculation steps to typical engineering design problems encountered by Chemical Engineering students that are imperfect. In other words, the engineering design presented by the QCT software for any given design problem would contain weaknesses, flaws or even errors. Students utilized the software to practice identifying weaknesses, flaws and errors in design calculations and then derived better or correct designs by applying concepts and knowledge acquired in class. The extent of achievement of the intended learning outcome was measured quantitatively by comparisons of performances in the final examination between students who have and those who have not utilized the QCT software to apply a repeated practice approach towards their learning of quantitative critical thinking. A positive correlation between frequency of usage of the software and performance in the quantitative critical thinking question in the final examination was observed, indicating that the software had been effective in improving this learning outcome.

600863 Measuring Student Academic Motivation Toward Process Safety Decision Making within a Virtual Digital Environment

Abstract:

In recent years our research team has developed an engaging digital environment or "digital game" we call *Contents Under Pressure.* In the game students are able to make decisions, many related to process safety, that arise when working as a practicing chemical engineer in industry. This immersive approach is a valuable strategy toward evaluating an individual's decision making ability and thought processes related to process safety and engineering ethics, where real-world incentives and disincentives are difficult or impossible to achieve in survey or classroom-type environments. A goal of this novel approach toward process safety education is to improve upon student academic motivation to learn process safety content, which we plan on assessing using the MUSIC model developed by Dr. Brett Jones. Student academic motivation may differ based on student gender identity, ethnicity, and perception of being a "gamer". The effect of these factors on student academic motivation as it pertains to interaction with *Contents Under Pressure* will be evaluated by the MUSIC model and presented in this work.

597857 Using Virtual Reality to Gain Deeper Insights into Transport Phenomena in Process Engineering

Abstract:

Students usually encounter transport phenomena in university teaching in the form of dimensionless numbers, e.g. to be read from diagrams. However, in process engineering apparatuses and reactors they are very complex, three-dimensional and under certain conditions transient. Furthermore, they depend strongly on the underlying design. With the help of Computational Fluid Dynamics (CFD), the occurring flow and associated transport phenomena can be modeled and simulated in great detail. In the master course "CFD for Process Engineering" at Clausthal University of Technology, students apply this modeling approach to investigate a current and practical problem from process engineering and to carry out their own digital design studies.

We set the aim is to apply, test and, if necessary, further develop new teaching and examination formats for this course. The focus here is on the integration of virtual reality (VR), see Fig. 1, into research-based learning. This technique enables students to achieve a previously unattained depth of immersion in their group work phase. Trained VR tutors also support the students with video tutorials, presentation techniques and individual supervision. In this contribution, we share recent experiences applying VR in this master course and discuss pros and cons of this new approach to active learning.

Fig. 1: Application of virtual reality in the master course CFD for Process Engineering

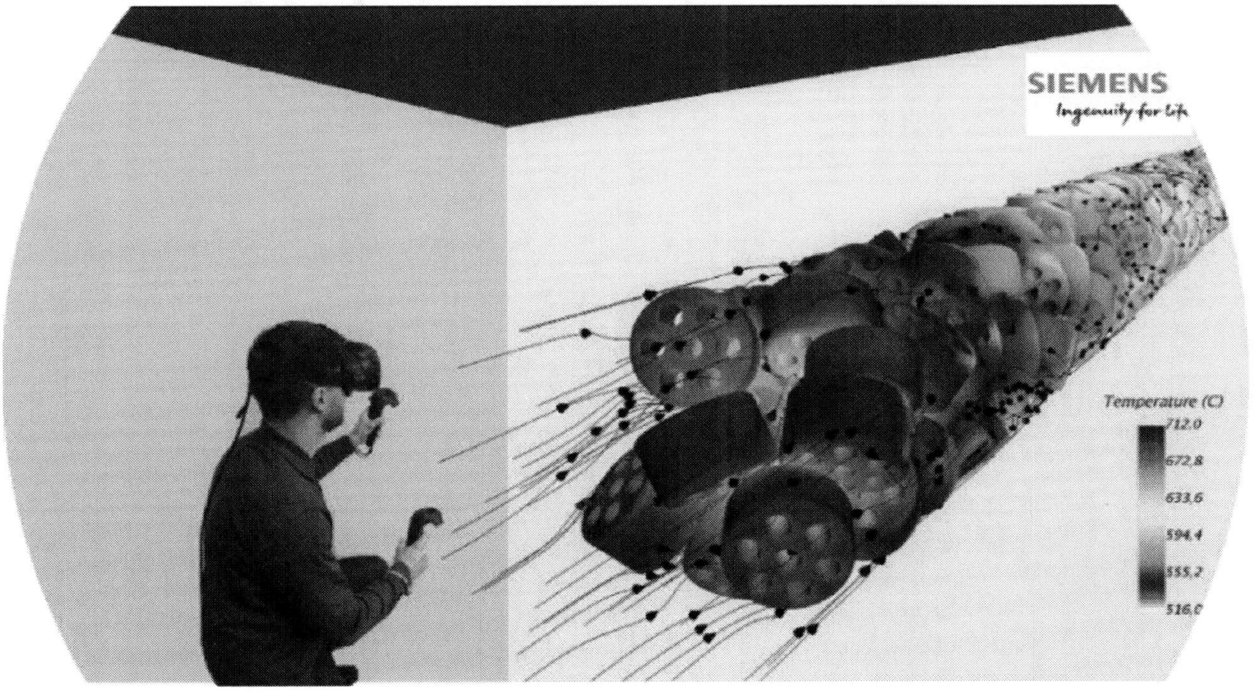

592747 Bringing ChemE-Sports, a Safety and Optimization Competition, to the Chemical Engineering Profession

Abstract:

SURG ChemE-Sports is quickly becoming part of the chemical engineering culture. Inspired by Esports, in 2019, an academic and industry team of a faculty member at Columbia University Chemical Engineering and Simulation Solutions Inc. developed a new method of using simulators and sponsored a national competition including 13 universities. The competition was created around the safety and optimization of a pentane-hexane distillation simulation owned by Simulation Solutions Inc. An NCAA Regional bracket was developed for the competition at the annual AIChE Meeting in Orlando, Florida on 10 November 2019. The results of the 2019 competition and lessons learned are presented here. This presentation will focus on the competition and how it was developed.

608806 The Strategy and Implementation of Sustainable and Green Chemistry Education in Taiwan

Abstract:

The Toxic and Chemical Substances Bureau (TCSB), to achieve a better environmental quality and reduce population health risk, start to promote general education of green chemistry (GC) in advanced education system, provide the research funding to academic research on GC, and enhance the partnerships with cross discipline sector, institutes, such as Department of Education and Ministry of Economic Affairs, and industries in Taiwan since 2017.

Based on international green chemistry online courses, experimental methods, and recommendation from local experts, TCSB has developed and proposed a syllabus and a series of online courses for general education of green chemistry in advanced education. At least 12 courses materials, including Introduction to 12 principals of green chemistry, environmental medicine, chemical substances safety, experiment for identifying plastics, life cycle assessment, and industrial ecology and so on, were established and revised by the editorial committee, and available on the TCSB website.

In conjunction with the domestic industries, we visited 30 companies (including chemical substances manufacture, recycling of recycled materials, paper industries, and electronics manufacturing) and conducted the questionnaires to explore the implementation of GC and their needs about labors' working capability developed from advanced education. Production and process management are most important background knowledge they need. In November of 2018, we held 4 GC general education workshops for college students (n=230) and 2 workshops for GC training for elementary school teachers (n=54). Interesting lectures, such as Definition of Green Chemistry and Introduction to Environmental Susceptibility, Relationship between Chemicals and Environment, Toxicity and Waste Reduction, Green Chemistry Circular Economics and so on were presented.

To provide a platform for college students to join the GC education activities, the Green chemistry Creative Competition for nationwide colleges and universities were initiated in 2017 and the second event will take place in 2020.

609595 Introducing Sustainability Criteria in the Chemical Engineering Design Projects: A Case Study of Green Ammonia

Abstract:

Sustainability has become a major concern for chemical companies seeking to remain competitive in the market place. Our graduates in chemical engineering must, therefore, be well equipped to address the challenge of designing processes to meet the growing demand for chemicals and fuels more sustainably.

In this talk, we would like to share with the audience our current efforts at the Institute for Chemical and Bioengineering (ICB) at ETH for including sustainability criteria in the design projects taught in the Chemical Engineering studies. Our approach is based on three main interconnected elements: (i) the choice of a highly relevant project in the area of sustainable fuels and chemicals; (ii) the use of computer-aided tools to carry out the necessary calculations to quantify the economic and environmental performance of a process; and (iii) the application of life cycle assessment (LCA) principles to measure the environmental footprint of a chemical considering its entire life cycle.

Organized in groups, the students work intensively on a given topic during one week to develop a preliminary process flowsheet and carry out mass and energy balances, as well as economic and environmental calculations. Daily meetings ensure that they progress smoothly towards the final goal of producing a sensible design meeting the goal specified. Sustainability criteria are embedded in the problem statement right from the very beginning, rather than considered "*a posteriori*" in the later stages of the process design.

Using green ammonia as a case study, we will explain how we define the project and motivate the students to find more sustainable production pathways for an existing chemical. The goal here is to replace fossil ammonia by green ammonia, where the later uses electrolytic hydrogen produced from renewable energy rather than fossil hydrogen generated from natural gas.

Overall, the focus on sustainability improves the students' experience by boosting their motivation while exposing them to the latest trends in sustainability.

607674 Product [X] –an Ongoing Experiment in Chemical Engineering Sustainability Education Using Project-Based Learning

Abstract:

In our daily lives, consumers interact with chemical engineering originated products such as the drinking straw, plastic cups, grocery bags, water bottles and others. These single-use products create waste problems. Eliminating or replacing these products with a sustainable solution is a complex task. It requires addressing consumer behavior, regulations, cost, mass-production, safety, health and environmental footprint to name a few. Different disciplines are needed to propose a holistic solution from engineering, business, health and social science perspectives. Traditional chemical engineering education focuses on math, science and engineering core theories where interactions with direct consumer-based products are limited. Moreover project-based classes in the traditional curriculum take place in the senior design capstone, which most often assigns to large scale chemical and industrial processes. Product [X] is an ongoing educational experiment in the chemical engineering department at Texas A&M University at Qatar. The project attempts to prepare students to deal with commercial consumer products as part of multidisciplinary teams. The aim is to promote 'circular' thinking in development and consumption, directly relating to the U.N. Sustainable Goals (Goal 12). At the same time, this project will motivate students to acquire and apply new knowledge outside of the classroom, thus contributing to a holistic engineering experience. Undergraduate chemical engineering students from all levels (freshman to seniors) are encouraged to work in teams with members from different majors to find solutions under the supervision of a chemical engineering faculty member. This student-led-project lasted over two semesters. In the academic year 2019-2020, the challenge was to redesign the ordinary plastic drinking straw, which was invented to eliminate a public health problem and ended up contributing to the plastic waste problem. A work plan is divided into two stages. The first stage started by short pitch presentations to a technical committee, which was followed by the implementation and execution of that plan. The faculty mentors helped in defining scope, feasibility, organizing workshops and connecting teams with stakeholders. The interactions were through face to face and virtual meetings. Currently, two teams selected to work on environmentally friendly prototypes using bio-based raw material and egg-shells. They investigate the synthesis and manufacturing of sustainable, feasible routes in the Middle East. A third team decided to pursue another direction to work on policy related to the case study. Different policy levels are explored from the university campus up to global ones. The planned impact of this educational experiment is to: fill a gap in the study curriculum, encourage self-learning outside the classroom, grow a network beyond the university campus, and encourages innovation through research or entrepreneurship. The students will apply engineering knowledge to define an open-ended problem, identify scale of the proposed solution, assess sustainability of the solution from environmental, social and economic perspectives.

607487 Hands-on Activities for Introducing High School Students to Alternative Energy and Climate Change

Abstract:

Alternative Energy and Climate Change are currently popular and commonly discussed topics of great world-wide significance to chemical engineers, practitioners, and the greater community of scientists. In an effort to engage local high school students from the surrounding communities of the New York City metropolitan area, a six-week hands-on and project-based approach to learning about the current challenges of renewable energy systems and climate change was developed as part of a Summer STEM Program at The Cooper Union. Through the design of experiments, small lectures, team-based activities, invited speakers, and field trips to nearby energy plants, students were introduced to the field of chemical engineering through the lens of renewable energy systems and climate change. In particular, a multi-week project involving different engineering aspects of biofuels production was included as a main focal point of the summer program.

Furthermore, this research-based opportunity provided students with activities to enrich their fundamental understanding of science, math, computational, and experimental skills. The students were also introduced to important economic, environmental, and global contexts and topics related to renewable energy and climate change. Examples of experiments performed during the program include testing the viscosity of oils and determining the specific heat of different substances. By studying these and other hands-on characterization techniques, students were able to apply this knowledge to the synthesis and characterization of biofuel derived from clean and waste resteraunt canola oil.

597120 Incorporation of Sustainable Ammonia Synthesis Process Design with Enhanced Assessment Method into Chemical Engineering Senior Design Course

Abstract:

Sustainability has become a key subject in chemical engineering (ChE) practice, largely due to the challenges such as natural resource depletion, greenhouse gas emission, economic globalization, increasingly stringent environmental regulation, etc. To incorporate the concept into ChE education, the Accreditation Board for Engineering and Technology (ABET) has listed sustainability as a crucial element in engineering curricula and stated that "Engineering programs must demonstrate that their students attain an ability to design a system, component, or process to meet desired needs within realistic constraints such as economic, environmental, social, political, ethical, health and safety, manufacturability, and sustainability". The senior capstone process design course is a key educational component to fulfill the goal, as it could incorporate subjects such as sustainable resources, biomass, energy efficiency, design standards and regulations, process safety, etc.

In this work, three ammonia synthesis process technologies are designed in the ChE 4451 - Process Synthesis and Design II, including (1) traditional Haber–Bosch ammonia synthesis, (2) large scale water electrolysis for ammonia synthesis, and (3) distributed wind-based water electrolysis for ammonia synthesis. Besides traditional design methodologies such as concept design, process simulation, heat integration, equipment sizing, and cost estimation, the new sustainable process requires enhanced assessment method to incorporate emission penalty, renewable resource credit, transportation cost, modular manufacturing discount, as well as safety/control concerns. The quantitative results show that if only traditional assessment is applied, the wind energy based distributed process could not compete with the traditional one. However, the new assessment method shows that the new process could have a competitive breakeven price when all the evaluation factors are considered.

600835 Teaching Sustainability As a Complex Systems Approach

Abstract:

Many problems of the present such as the need for water, food, and energy have first been attempted to be solved through one-sided solutions, e.g. intensive agricultures or exploiting fossil fuels.[1] As identified by Rittel and Webber (1973) however, such a class of social-ecological global dilemmas should be considered as *wicked problems*.[2] Wicked problems are complex issues, which lack clear definition and which may not be solved through traditional modes of decision-making. One may want to contrast them to *tame problems,* for which our scientific knowledge and common procedures have been developed so far.[1,2] Working with such problems requires, amongst others, an education that equips students to handle cross-disciplinary and macro-level thinking,[3]with the ability to handle interconnections and dependencies, and to work with dynamic rather than static structures.[4] Fenner et al.[5] follow a similar line of argumentation when outlining that within engineering education, understanding complexity and its relationships should be the starting point as sustainability problems require a system view of the world. Altogether, educators must find a different educational paradigm, which prepares students for the new challenges faced,[6] not only in engineering but in every academic discipline.[7]

One of the main challenges faced by the educator is finding appropriate means (terminology as well as methodological approaches) that allow interdisciplinary work, discussions, and communication. We argue that sustainability problems must be regarded as complex systems and indicate that they should be studied with methodologies allowing for the complexity. One well-established approach to handle complexity is through Network Science. Within Network Science, real-world systems are represented by two basic units; nodes commonly represent any object and/or subject in the system, and edges, that represent relationships. A street network, the brain, water distribution, energy grids, and the internet – we recognise networks everywhere.[8]Network Science offers a common language for scientists to deal with complexity especially when experts from different areas are needed. Besides a common language, it provides both a mathematical formalism and data driven computational tools to quantify aspects and behaviour of a complex system.[9]

Herein, we introduce an interdisciplinary workshop outline which was developed as course material for an extracurricular teaching activity for a group of 17 participants during a three-day full-time workshop. We explain the basics of Network Science and its terminology and how we guide the participants towards using this analysis for a sustainability case study. The participants were asked to identify and characterise synergies and trade-offs between the United Nations (UN) Sustainable Development Goals (SDGs) by using network metrics. We recognised the importance of interdisciplinarity because both Network Science as a tool and also its simplifications were thoroughly discussed between the disciplines. The participants defined a synergetic relationship as a link in the network and discussed circular structures where the enhancement of one goal leads to a theoretically infinite enhancement of a substructure. Previous scientific work [10,11] was compared and discussed with the ideas and models of the participants at the end of the sessions. Feedback possibilities during the course and a reflection forum at the very end led to a qualitative assessment of the teaching method; it showed high contentment both with learning more about sustainability and learning a new methodological approach in an interdisciplinary group.

This work introduces a novel methodological approach in order to understand, discuss, and critically evaluate sustainability problems and their complexity. It offers a teaching possibility which bridges gaps between disciplines, and which has the potential to contribute towards solving most challenging questions of this century.

Acknowledgments:

JMW and CPL gratefully acknowledge the German National Scholarship Foundation as funding body for the seminar as well as Tim Deisemann and Lisa Oswald as co-organisers. They thank all participants of the workshop, who made their ideas become reality, and all other lecturers, who have committed much time and effort for similar course concepts. CPL acknowledges Goldbeck Solar for all their support and JMW the Department of Chemical Engineering for funding of her PhD studentship.

References:

1. Brown, V. A. *et al. Tackling wicked problems through the transdisciplinary imagination.* (Earthscan, 2010).
2. Rittel, H. W. J. & Webber, M. M. Dilemmas in a General Theory of Planning. *Policy Sci.* **4**, 155–169 (1973).
3. Seager, T., Selinger, E. & Wiek, A. Sustainable Engineering Science for Resolving Wicked Problems. *J. Agric. Environ. Ethics* **25**, 467–484 (2012).
4. Warburton, K. Deep learning and education for sustainability. *Int. J. Sustain. High. Educ.* **4**, 44–56 (2003).
5. Fenner, R. A., Ainger, C. M., Cruickshank, H. J. & Guthrie, P. M. Embedding sustainable development at Cambridge University Engineering Department. *Int. J. Sustain. High. Educ.* **6**, 229–241 (2005).
6. Davidson, C. I. *et al.* Preparing future engineers for challenges of the 21st century: Sustainable engineering. *J. Clean. Prod.***18**, 698–701 (2010).
7. Parker, K. A. *et al. Teaching Sustainability/ Teaching sustainably*. (Stylus Publishing, 2012).
8. Barabási, A.-L. & Bonabeau, E. Scale-Free Networks. *Sci. Am.* **288**, 60–69 (2003).
9. Barabási, A.-L. *Network science*. (Cambridge Univeristy Press, 2016).
10. Le Blanc, D. Towards Integration at Last? The Sustainable Development Goals as a Network of Targets. *Sustain. Dev.* **23**, 176–187 (2015).
11. Stockholm Resiliene Centre. *Stockholm Resilience Centre's (SRC) contribution to the 2016 Swedish 2030 Agenda HLPF report*. (2017).

598622 Development of Middle School STEM Classroom Lesson Plans and Afterschool Program Activities to Support USDA-Sponsored Project on Alternative Crop Bioenconomy

Abstract:

The Sustainable Bioeconomy for Arid Regions (SBAR) project is a USDA-funded coordinated agricultural project (CAP) between the University of Arizona, Colorado School of Mines, Colorado State University, New Mexico State University, Bridgestone Americas, and the USDA Agricultural Research Service. SBAR focuses on development of the feedstock, processing, and logistics of two alternative crops for the desert Southwest: guar (*Cyamopsis tetragonobola*) and guayule (*Parthenium argentatum*). As one of nine USDA regional biofuels and bioproduct CAPs, SBAR includes research, education, extension, and outreach components. SBAR research components include plant breeding and agronomic trials, value-added co-product development, and sustainability modeling. The extension and outreach components include demonstration plots, farmer workshops, extension publications, and 4-H youth camps and programs.

The SBAR education component focuses on workforce development to enable suitable people to fulfill the jobs needed in the developed bioeconomy, specifically to create and disseminated STEM education materials at the middle school level. Graduate student fellows (4 in NM and 6 in AZ) were paired with public middle school teachers to draft and vet classroom and afterschool program activities and materials. Prior to the school year, fellow/teacher pairs attended a week-long science concept orientation, followed by a week of professional development to translate the STEM concepts into suitable lesson plans for middle school students. Fellows, mostly engineering and science majors, assisted in the classroom for at least 10 hrs./wk. to develop their own communication and teaching skills. Deliverables for fellows included a working draft of an SBAR-related lesson plan, with input from their mentor teachers, and formal/informal format hands-on activities. Two of the SBAR fellows worked with their teacher mentors to start an after-school program called "**Guar**dians of the Biosphere" (GOB). Activities for GOB to date have included food calorimetry, biofuel synthesis, giant bubbles (surfactants), and aquaponic agriculture, all with discussions of laboratory safety and engineering principles.

608410 Active Membranes for Neuron-Inspired Ionotronic Devices

Abstract:

The membrane of a neuron recognizes chemical signals at the synapse, converts them to electrical signals that move rapidly down the axon upon meeting a threshold, and finally triggers the release of another chemical signal at the axon terminal. Assemblies of these building blocks – nervous systems – comprise the most sophisticated computing schemes yet known and make possible the autonomous function and rich behavioral diversity of all higher animals.

With the intent of enabling advanced materials with complex autonomous functionality, we are developing a gel-based scheme inspired by the architecture of neurons that converts external chemical and optical signals to fast-moving electrical potentials, which can then be harnessed to perform chemical work. By storing energy in a chemical gradient between aqueous compartments and releasing it by triggering selective permeabilization events in an initially insulating organogel membrane, we can generate electrical potential changes on the order of 100 mV. We compare the spatial voltage decrement along insulating, capacitive membranes to that of an electrotonus in excitable tissues. Finally, we describe an electrochemical end effector whose activity is initiated by the electrical impulse. In this way, we aim to exploit the toolkit of stimuli-responsive chemistry to enable fast, spatially decoupled responses to a variety of environmental signals.

611348 Electrochemical Switching of a Fluorescent Molecular Rotor Embedded within a Bistable Rotaxane.

Abstract:

I report how the constrained environment introduced by the mechanical bonds within an electrochemically switchable bistable [2]rotaxane controls the rotation of a fluorescent rotor, namely 8-phenyl-substituted boron dipyrromethene (BODIPY). The electrochemical switching of the bistable [2]rotaxane induces changes in the ground-state co-conformation and the corresponding excited-state properties of the BODIPY rotor. In the starting redox state, when no external potential is applied, the cyclobis(paraquat-*p*-phenylene) ($CBPQT^{4+}$) ring component encircles the tetrathiafulvalene (TTF) unit on the dumbbell component, leaving the BODIPY rotor unhindered and exhibiting low fluorescence. Upon oxidation of the TTF unit to a TTF^{2+} dication the $CBPQT^{4+}$ ring is forced toward the molecular rotor leading to an increased energy barrier for the excited-state to rotate the rotor into the state with the high non-radiative rate constant, resulting in an overall 3.4-fold fluorescent enhancement. On the other hand, when the solvent polarity is high enough to stabilize the excited charge transfer state between the BODIPY rotor and the $CBPQT^{4+}$ ring, the movement of the ring towards the BODIPY rotor produces a strong fluorescent signal decrease. This fluorescence quenching results from photoinduced electron transfer from the BODIPY rotor to the $CBPQT^{4+}$ ring, as demonstrated by femtosecond transient absorption spectroscopy. The modulation of switchable electrochromic behavior makes this bistable [2]rotaxane amenable to electro-optical applications.

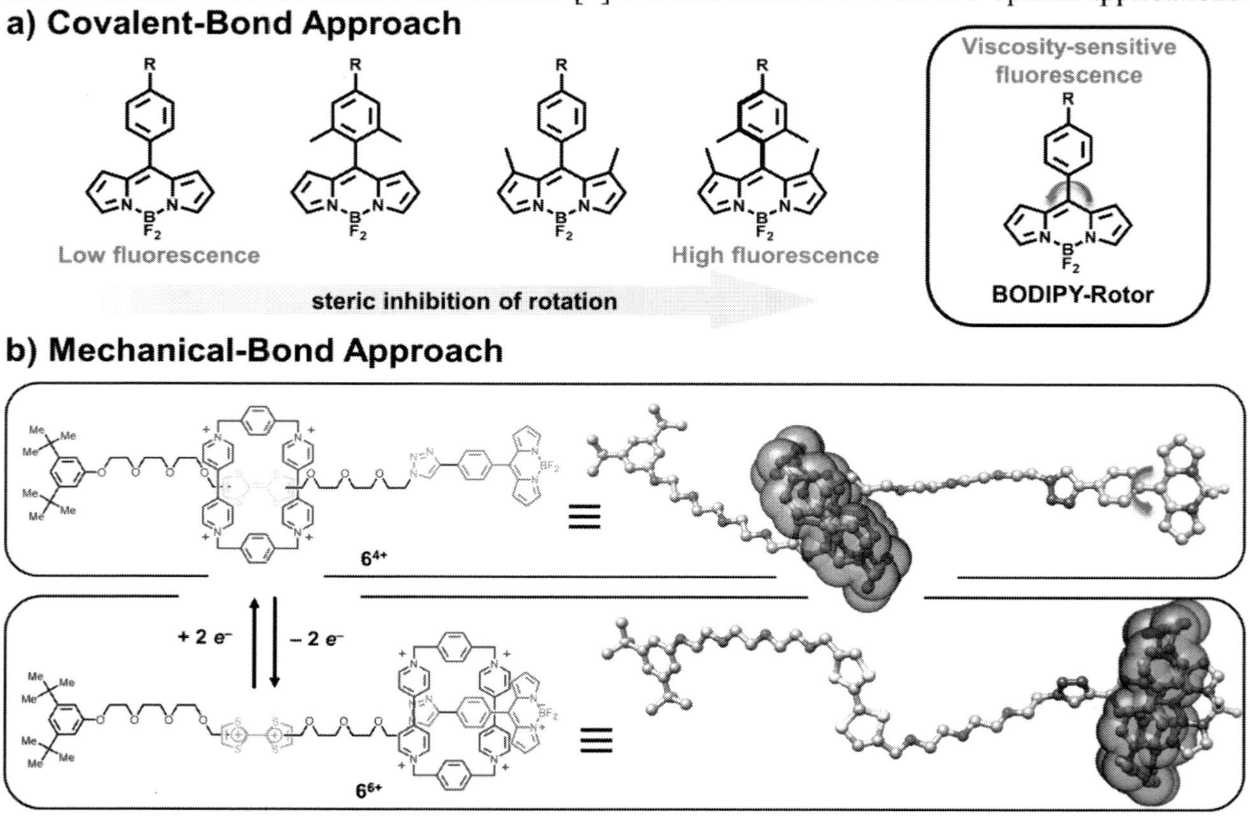

606952 Green Synthesis of Zinc Sponge Opens Sustainable Battery Supply Chains

Abstract:

Scientists have devoted significant efforts highlighting the need for batteries that enable the use of sustainable energy sources such as wind and solar but have paid much less attention to the need for sustainable batteries. We seek to identify sustainable, rechargeable batteries and to create simplified, high-performance, low-cost battery electrodes. To identify promising candidates, we review supply-risk data and the abundance of battery-relevant elements in the Earth's upper continental crust. We find that of the rechargeable-battery chemistries that rely on abundant low-risk elements, zinc (Zn) batteries, namely Zn–MnO_2 and Zn–air, are the lowest cost and most energy dense. Guided by this analysis, we advance Zn batteries by reporting a green synthesis that lowers the cost of rechargeable Zn-sponge electrodes by 74%. We test these electrodes in Zn–air cells and demonstrate a 3,200% increase in rechargeable areal capacity over the majority of previously reported work.

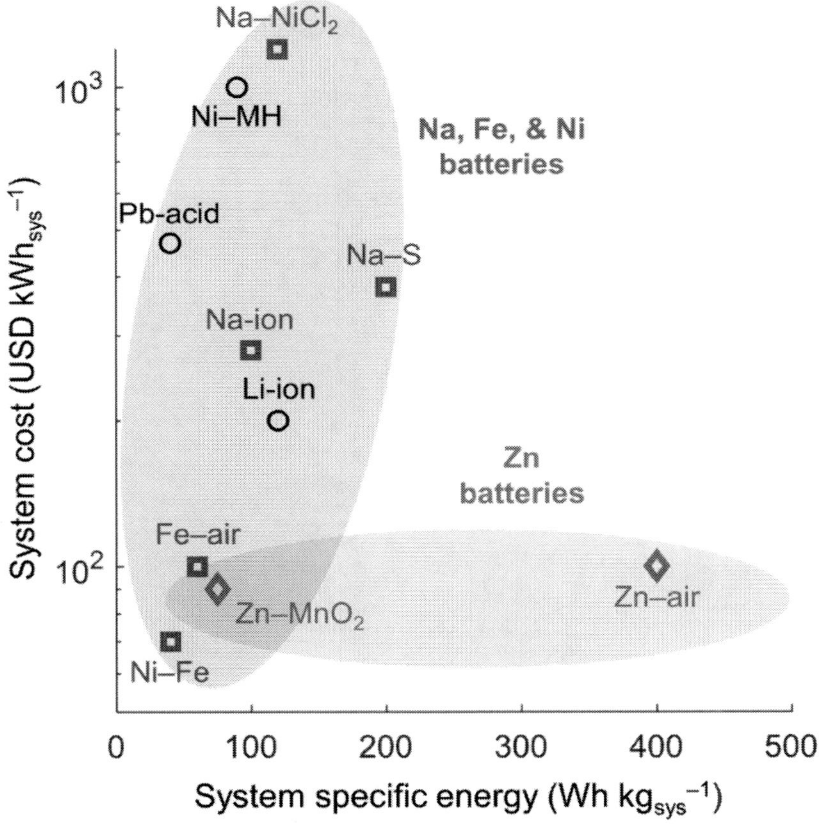

602410 Optimized Mesostructures of Li-Ion Battery Electrodes Predicted from Particle-Based Simulations

Abstract:

Electrochemical performance of Li-ion battery electrodes is crucially governed by the morphology of carbon-binder domain (CBD)—composed of carbon black nanoparticles and polymeric binder—in the local vicinity of electroactive particles such as $LiNi_{1/3}Mn_{1/3}Co_{1/3}O_2$. However, the effect of CBD properties, such as its cohesive strength and adhesive strength between CBD and electroactive particles, on electrode mesostructure is not well understood. This results primarily from our inability to visualize mesostructural details at sufficiently low resolution and large domain sizes within experiments, and numerical challenges associated with accurately modeling CBD topology within the complex and tortuous electrode mesostructure. This talk will describe our physics-based high-fidelity models to simulate the dynamics of micron-sized electroactive particles and nanometric CBD particles during various stages of electrode fabrication. Several experimentally-relevant compositions of CBD, electroactive material, and electrode porosities will be considered. The effect of cohesive and adhesive strength of CBD on its local topology around electroactive particles and on the porous mesostructure will be discussed. Lastly, finite element modeling of electronic and ionic conductivities of these simulated electrode mesostructures will be presented, and predictions towards mesostructure optimization for enhanced electrochemical properties will be discussed.

610317 Understanding Spatial Heterogeneities in Ion Transport in Composite Solid-Ion Conductors Based on Neutral Plastic Crystal–Polymer Hybrids

Abstract:

Heterogeneities intrinsic to composite solid-ion conductors (SICs) are thought to influence the solvation structure and dynamics for lithium ions (Li^+) responsible for the ion current, as well as the diffusion paths available for ion transport. An understanding of the molecular origin of such heterogeneity and its impact on bulk Li^+ transport is needed to intentionally design a SIC showcasing desirable properties for integration into a lithium metal battery. Here, we show that a lithium halide doped plastic crystal–polymer composite naturally forms a defect rich, high lithium mobility domain around the polymer backbone that is correlated with higher frequency of bond rotation within the plastic crystal. We show that a new class of SIC delivers high conductivity (1–5 mS cm^{-1} at T = 20–25 °C), high transference number (t^+_{SS} = 0.72) and naturally protects against overcharging in lithium metal battery. The key to our success is to engineer a defect rich domain through which the lithium ion diffusivity is order of magnitude higher than the bulk plastic crystal. Our choice of lithium halide also forms an electronically insulating interlayer in lithium metal battery allowing lithium to be reversibly plated at area specific resistance of 60 ohm cm^2 for >300 h. We anticipate that our work will reinvigorate the field of plastic crystal-based SICs as a new design principle for tailoring their transport and mechanical properties for use in solid-state batteries.

602254 Mechanistic Understanding of Electrochemical Processes in Alkaline Environments

Abstract:

Alkaline fuel cells (FCs) have attracted attention for potential applications in efficient energy production. The higher stability of non-precious metals in alkaline environment offers the possibility to synthesize cheaper electrocatalysts, lowering the overall cost of FCs [1]. However, non-precious metal anodic electrocatalysts are considerably less active than more commonly used PGM-electrocatalysts. As a result, successful wide-scale development of high-performance alkaline precious metal-free FC remains a challenge.

Mechanistic understanding of electrocatalytic processes could greatly contribute to the design of improved electrocatalysts. At present, comprehensive mechanistic models capable of capturing the *in situ* nature of the catalytic active site in alkaline environment are still lacking for cases such as the oxygen reduction reaction (ORR) and the ammonia oxidation reaction (AOR).

In this contribution, we utilize electronic-structure calculations in the framework of coverage-cognizant DFT models to unravel complex reaction mechanisms for ORR and AOR in alkaline environment. We will focus on ORR on Au(100) and AOR on Pt(111), with two key questions in mind: i. "*What is the origin of the remarkable activity of Au(100) for ORR in alkaline environment?*"[2] and ii. "*What is source of Pt deactivation in the AOR?*"[3-6].

[1] Firouzjaie H. A., Mustain W. E., *ACS Catal.*, **10**, 225-234 (2020)

[2] Staszak-Jirkovský J, Subbaraman R., Strmcnik D., Harrison K., Diesendruck C. E., Assary R. Frank O., Kobr L., Wiberg G., Genorio B., Connell J.G., Lopes P.P., Stamenkovic, V.R., Curtiss L., Moore J., Zavadil K.R., Markovic N. M., *ACS Catal.*, **5**, 6600-6607 (2015)

[3] Rosca V., Koper M. T. M., *Phys. Chem. Chem. Phys.*, **8**, 2513–2524 (2006)

[4] Matsui T., Suzuki S., Katayama Y., Yamauchi K., Okanishi T., Muroyama H., Eguchi K., *Langmuir*, **31**, 11717–11723 (2015)

[5] Vidal-Iglesia F. J., Solla-Gullón J., Pérez J. M., Aldaz A., *Electrochem. Commun.*, **8**, 102–106 (2006)

[6] Herron J. A., Ferrin P., Mavrikakis M., *J. Phys. Chem. C*, **119**, 14692–14701 (2015)

601551 New Frontiers for Electrochemistry in Addressing Climate Change: Carbon Capture and Storage

Abstract:

The coronavirus pandemic in early 2020 cascaded a frightening threat, which upended familiar routines, disturbed the global economy, and endangered people's lives around the world. Despite scientists long warned this might happen, a little was actually done before the pandemic to prevent the destructive consequences. A similar warning is being repeatedly echoed through the voice of scientists regarding the climate change and its multiscale devastating impacts on the planet and human's life. Designing any action plan to mitigate climate change must include a detail strategy to lower greenhouse gas emissions especially carbon dioxide (CO_2) as repeatedly cited to have the most significant contribution to the observed increase in global average temperature since the mid-20th century.

Electrochemistry, a key part of the 2019 Nobel Prize in Chemistry, is believed to be a powerful tool for designing diverse CO_2 mitigation approaches, offering a unique advantage in the future, when the electrical energy is mostly generated by renewable; the carbon footprint and the cost of the electrochemical processes would be minimized. In this context, electrochemically driven processes for carbon capture and storage (CCS), as a new frontier within the electrochemistry and climate change framework that has drawn significant attention in the last few years, could be an attractive approach to either avoid further release of CO_2 by capturing it at the source point or to reduce the atmospheric CO_2 level through a direct air capture process.

In this talk, two recently developed electrochemical based processes for CCS, namely proton concentration process (PCP) and electrochemically mediated amine regeneration (EMAR), will be discussed and recent advancements in process developments and scale up will be provided. In short, in a PCP, the pH sensitivity of CO_2 hydration is leveraged such that CO_2 is absorbed as bicarbonate and carbonate ions at high pH values and desorbed as gas at low pH values in an electrochemical cell with proton intercalating electrodes (Figure 1A). The EMAR process scheme is similar to that of the state-of-the-art amine-based thermal scrubbing processes, but the high-temperature desorber (normally operating at 120 ^{o}C) is replaced by an electrochemical cell that operates at moderate temperatures (e.g., 50 ^{o}C) in which copper ions are electrochemically generated from a copper plate anode to drive the dissociation of amine and CO_2 toward CO_2 desorption. Once the gas is flashed off, the CO_2 lean stream is regenerated via the electrochemical plating of copper on the cathode from the copper-amine complex (Figure 1B). For both PCP and EMAR, the process scheme consists of three chemical or electrochemical transitions that take place in an absorber, an anode, and a cathode. The electrochemical work requirement for the PCP and EMAR to desorb CO_2 captured from a flue gas stream are estimated to be 33.2 and 31.3 kJe/mol$_{CO2}$, respectively, suggesting the competitiveness of these approaches. These electrochemical based technologies offer several advantages over thermally driven amine-based processes, including operation at low temperature and the ability to desorb CO_2 at moderate pressures, which minimizes the downstream compression costs for CO_2 storage. Overall, the developed PCP and EMAR could be impactful electrochemical-based technologies that could effectively mitigate CO_2 emission and climate change.

Figure 1. Schematics of (A) a proton concentration process (PCP) and (B) an electrochemically mediated amine regeneration (EMAR) developed for carbon capture from a flue gas stream. For both PCP and EMAR, the process scheme consists of three chemical or electrochemical transitions that take place in an absorber, an anode, and a cathode.

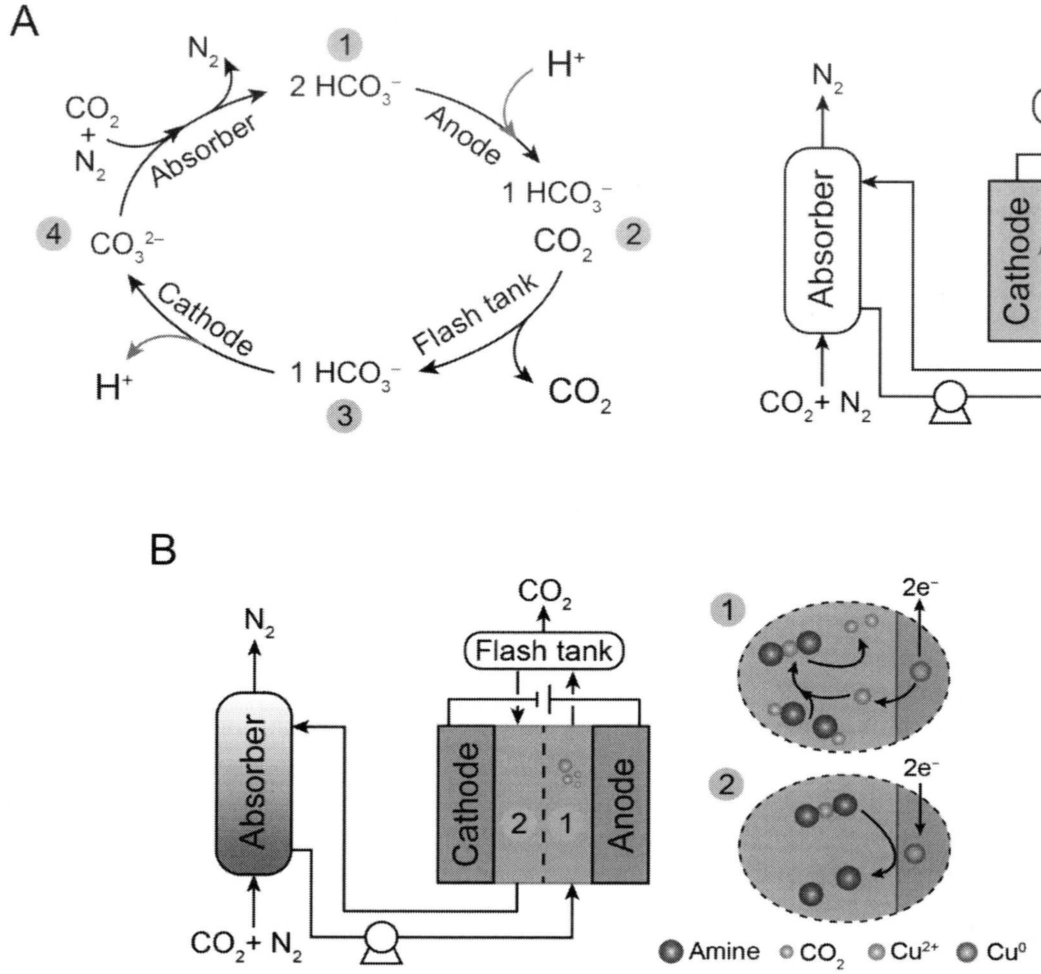

611731 Data-Driven Modeling of Electrochemical Systems

Abstract:

A convergent stream of big data, modeling and analytics in electrochemical systems enabled by high fidelity and throughput experimentation and computation creates a unique opportunity for the discovery of novel chemical physics and engineering guidance. In this talk, I will present work on learning from microscopic images, diffraction and atomistic simulations. Examples include 1) quantifying previously inaccessible or inaccurate knowledge of the electrochemical reaction kinetics, free energy, surface kinetic map and chemo-mechanical coupling from high dimensional images and videos of reacting or equilibrated lithium iron phosphate (LFP) particles (3D, x,y and time, or even 4D x,y,k_x, and k_y), 2) extracting autocatalytic reaction kinetics of lithium in nickel manganese cobalt oxide (NMC) out of in-situ X-ray diffraction and electroanalytical methods, and 3) coarsening of molecular simulations, all of which achieved quantitative matching between data and model. These applications show new ways in which experimental and simulation data can be extremely useful in revealing mechanistic insights about fundamental electrochemical processes (such as the suppression and induction of phase separation) and generating knowledge that can be translated into making new predictions through continuum modeling. I will introduce the modeling, inversion and statistical techniques that lie at the core of this endeavor.

607128 Scalable, Hands-on, Problem-Based Learning in Unit Operations Lab to Improve Student Problem-Solving Skills

Abstract:

The benefits of hands-on, problem-based learning have been well documented in the literature, but implementation of such pedagogical practice is often difficult in typical chemical engineering curricula. The origins of the difficulty include institutional pressure to reduce time to graduation, lack of space, cost, large enrollments, and safety concerns. This presentation explains how a safe, hands-on, problem-based learning project was added to a typical senior-year Unit Operations Lab for moderate cost, without increasing credit hours, and in a way that is scalable to an arbitrary number of students.

The purposes of this lab are to give students practice solving a more open-ended problem than is typically found in the curriculum and to improve engineering intuition about real systems over what is obtained from book work or other labs. Concerning the former, the problems presented 1) require meeting multiple competing design constraints, 2) have many possible solutions, 3) allow real possibility for failure, 4) require learning new skills not taught in previous courses, and 5) require choices to be made at every stage of the process from initial planning to final reporting. Concerning the latter, students come to realize the skills and time needed by those who build real systems, gain firsthand knowledge of flow system behavior, and come to understand the interplay between different subsystems and pieces of equipment.

During this experience, student build a pilot-scale flow system involving pumps, pipes, flowmeters, valves, tanks, pressure transducers, etc. to meet specified design constraints. Multiple sizes and types of each piece of equipment are available, and students are required to satisfy the design specifications —which include flowrate maximums, costs, space constraints, total length of pipe, etc.—using these materials. There are many ways to satisfy the design goals, but students seek to do so using the most efficient system possible. Moreover, they must prove their choice is an optimal solution to the problem and share this proof in their written reports.

This presentation will describe the hands-on, problem-based learning project in more detail. It begins with an overview of the learning outcomes for the project. This is followed by a discussion of the problem statements presented to students and an explanation of the equipment and space available to the teams. Examples of student projects are then presented. Finally, survey-based data on the value of this pedagogical approach are presented. These surveys, taken by 85 students from 6 different sections taught since 2013, include responses of students obtained just after the class was taken and again after the students have taken the second half of Unit Operations Laboratory. Survey results are also presented from alumni (N=51) who took the course and are now working in industry or are in graduate school. All the data demonstrate the very high value that students place on the experience both during and after the course. Taken as a whole, the results reconfirm that hands-on, problem-based learning is a very effective pedagogical approach. More importantly, the work shows that the method of implementation is scalable and can be incorporated into most current curricula for moderate monetary and space costs.

610894 Excel-Based Process Economics Computational Tools for Senior Level Process Design Courses

Abstract:

Chemical engineering economic calculations comprise an indispensable component of the decision-making process for grass-root design and debottlenecking studies of manufacturing plants. Understanding economic and finance calculations and their impact on the profitability and sustainability of chemical plants in the design stage or during routine operations is very important aspect of the education of future chemical engineers.

Chemical engineering departments traditionally integrate economic and finance calculations into their curricula as a separate course or module which is covered in the senior design courses. Typically, key elements, including equipment capital and operational costs, Net Present Value (NPV) or Internal Rate of Return (IRR) calculations, depreciation, tax, and interest are covered in the process design courses. Students usually obtain equipment and operational costs from various cost estimating software tools that are available either free or through academic license agreements; these cost estimates are then used to perform miscellaneous finance calculations such as depreciation, IRR and NPV. Even though expenses generated by computer tools give students some baseline, it is not clear how these calculations are done and how accurate they are. Moreover, these tools have a significant learning curve, and are introduced to the students during the last year of their studies; furthermore as mentioned earlier they require some type of academic license for their use.

To overcome these challenges, a team of senior students from the Chemical Engineering Department of the University of Illinois at Chicago (UIC), under the guidance of two faculty members developed a Process Economics Calculator in Excel using the Visual Basic for Applications (VBA) programming language. Through a well-designed interface, users of the Process Economics Calculator select and enter the sizing parameters of the process equipment along with some operational details; once the selection of the equipment is completed, equipment purchase and installed costs as well as operational costs are automatically estimated, and finally NPV values are calculated using data from the book Chemical Engineering Design, 2nd Edition by Gavin Towler and Ray Sinnott (2013 Elsevier Ltd.). Sensitivity analysis studies can also be carried out to gauge the effect of key process variables. The ease and accessibility of this Excel-based tool constitute a significant step in teaching process design and economics. The tool can be easily incorporated in the early years of chemical engineering curricula. Using this economic tool, first, second and third-year level chemical engineering students can be introduced to engineering economics and how they impact the decision-making process. The earlier exposure to economics will make it easier to integrate engineering design elements into intermediate-level courses. By promoting this engaging and easily accessible tool, students can have the opportunity to become better decision-makers in engineering and better problem solvers.

In this presentation, we will give a detailed description of the Excel-Based Process Economic Calculator, describe its most important features and demonstrate it with a number of examples drawn from the process simulation, optimization and design courses at UIC.

609439 Student Confidence and Metacognition in a FE Review Course in Chemical Engineering

Abstract:

The Fundamentals of Engineering (FE) exam is a nationwide, criterion-referenced, discipline-specific test that engineering students typically take late in their senior year, most often as a first step toward engineering licensure as a Professional Engineer.

The present study focuses on chemical engineering (ChE) seniors who completed a 3-credit review course, in part to prepare for taking the FE exam, but also to review major ChE topics (material/energy balances, mass/heat transfer, fluid mechanics, reaction engineering, materials science, process control) in preparation for a capstone design course the following semester. The first nine weeks of the course were focused on reviewing FE material mostly by solving FE-type problems with a varying range of difficulty. The course instructor used active learning methods during lectures and discussion sections such as Think-Pair-Share and Skeleton Notes. Students obligatorily took a half-length FE practice exam, referred to in this paper as the mock FE exam, near mid-semester. Those who did not pass the practice test, where "passing" was identified as correctly answering 50 percent or more questions on the exam, took a second practice test at the end of the semester.

To track students' gains in ChE knowledge and problem solving, change in metacognitions and in attitudes, students submitted responses to weekly surveys for which they received a small homework credit. The questions focused on rating students' confidence before and after solving each problem set, their degree of reflection when solving problems, and their problem-solving strategies. Student mastery of the technical content was established based on topic-level performance assessment through the Mock FE exam.

Our study was designed to address the following research questions, using quantitative analytic methods: 1. Does students' confidence increase with the number of weeks of review—i.e., is there a cumulative effect for confidence? 2. Does students' confidence to solve FE problems change after reviewing FE review topics? 3. Is confidence correlated with the mock FE test score? 4. Do students become increasingly reflective with each additional week of review—i.e., is there a cumulative effect for reflection? 5. Is reflection correlated with the mock FE test score?

Using a mixed-methods research design, we applied quantitative-qualitative data analysis methods to report changes in students' perceived confidence, strategy use, and metacognition. Statistical correlations were computed between the survey responses and student performance on the FE practice exam. We consider the present findings in the context of existing literature in engineering education on students' confidence, strategies, and metacognitive practices.

Quantitative survey responses showed significant gains in confidence after FE topic review activities and relatively consistent benefits in FE test performance associated with confidence ratings and metacognitive reflection ratings. The strong correlations between confidence and problem-solving performance are consistent with the didactic admonitions of Wankat and Oreovitz [1] and the instructional practices implemented by Woods and colleagues [2] [3] in engineering education. The present study also demonstrates that ongoing collection of students' reactions across the semester has the potential of providing instructors with evidence-based motivation for keeping some teaching activities the same and for changing others, in order to make teaching most effective.

References

[1] Wankat, Phillip C., and F. S. Oreovicz. Teaching engineering. McGraw-Hill, 1993

[2] Woods, Donald R. "An evidence-based strategy for problem solving." Journal of Engineering Education 89.4 (2000): 443-459.

[3] Woods, Donald R., A. N. Hrymak, R. R. Marshall, P. E. Wood, C. M. Crowe, T. W. Hoffman, J. D. Wright, P. A. Taylor, K. A. Woodhouse, and C. G. Bouchard. "Developing problem solving skills: The McMaster problem solving program." Journal of Engineering Education 86.2 (1997): 75-91.

612017 A Review of Alternatives to Physical Chemical Engineering Laboratories

Abstract:

The authors present a summary of a literature review examining evidence of effectiveness and the range of availability of alternatives to on-site laboratory instruction in chemical engineering. Recent events have raised concerns over the ability of chemical engineering programs to adapt their laboratory courses to a scenario where on-site work is disallowed. While work over the end of the Spring 2020 terms have focused on "making it through", it is appropriate to thoughtfully consider how key learning objectives targeted by typical chemical engineering laboratory courses can be addressed by alternative means. Three key categories of off-site laboratories are considered: simulated laboratory exercises, take-home laboratory where program provide supplies and equipment, and at-home laboratories using commonly found household items are considered. Available evidence of effectiveness in the context of common course outcomes (as categorized by Feisel and Rosa, JEE, 2005) provides the framework for consideration.

611291 Conceptests for Process Controls Courses: By Students, for Students

Abstract:

ConcepTest questions are frequently used in chemical engineering instruction, in part because of their availability through learning repositories like LearnChemE.com and AIChE Concept Warehouse. ConcepTests evaluate the readers' understanding by integrating the concepts into questions that force the learners to evaluate their own arguments for certain multiple choice answers while selecting one final answer. Thus, the learners' considered reasoning before selecting an answer allows for evaluation of conceptual understanding. However, while there are extensive questions available for almost all core undergraduate chemical engineering courses, few ConcepTests have been developed for Process Control courses.

As an assignment in a Process Control course at Northeastern University, students were given the opportunity to write ConcepTest questions of their own for use with future iterations of students. Students had to develop multiple choice questions with detailed solutions, and were allowed to develop the ConcepTests with respect to any concepts in the course. More than 40 students over two semesters developed questions that were then used throughout the following semester. This study discusses the range of questions created, the benefit of having students develop their own questions, and their benefit in future semesters.

602233 Initial Analysis of the Effects of Homework Wrappers in a Problem-Solving Course

Abstract:

A previous study spanning 30 years examined the combined effects of (a) changing from lecture to flipped class course delivery, (b) changing to predominantly effort-based homework grading and (c) introducing homework wrappers as a required component of homework assignments. That study took place in a junior-level, problem-solving course, specifically a kinetics and reaction engineering course. Prior to these changes the data showed a weak effect of homework completion upon exams scores, and the exams scores distribution of students who completed an above-average amount of homework was statistically equivalent to that for students who did not. After these changes, the effect of homework completion upon exams scores more than doubled (from 0.171 to 0.435 exam points per percent homework submitted) and the average exams scores of students who submitted an above-average amount of homework was 12.45 points higher than students who did not.

In this study, the effects of homework wrappers are examined in greater detail. The rationale for introducing homework wrappers is that student learning is improved by practice coupled with targeted feedback. Grading a problem and returning it along with a correct solution allows students to understand *what* they did wrong. The intended purpose of homework wrappers in the present context is to cause the students to examine the thought processes they employed when solving the problem and identify and correct points where their approach was flawed or their understanding was incomplete. The previous study showed that wrapper use correlated with higher exams scores, but it did not examine whether the wrappers were actually providing this level of feedback to the students.

Prior to assigning a given type of problem to the students as homework, they were instructed on how to identify that problem type and a general approach for solving that type of problem was presented. A problem of that type was then solved as an in-class learning activity. The activity was scaffolded to emphasize the implementation of the generalized approach to solving that type of problem. Only then was a problem of that type assigned as homework. As implemented here, homework wrappers were follow-up assignments wherein students are asked reflect upon their solution to a problem, write an analysis of their thinking at the time they formulated their problem solution, comparing it to the general approach, and identify knowledge and skills they were missing, etc.

Kinetics data analysis problems were selected for study. These are one of several problem types wherein the students must formulate mole and energy balances for a reactor in order to solve the problem. This problem type involves parameter estimation using those balances and experimental data. Other problem types use the same balance equations, but in a different way. Kinetics data analysis is considered relatively early in the course. As a consequence, kinetics data analysis problems appear in at least one mid-term exam and again in the final exam.

Each homework assignment involving a kinetics data analysis problem was followed up with a homework wrapper. The student response on each wrapper was coded to indicate points that the student identified as a point where their thinking was flawed or the understanding was incomplete. Separately, each submitted assignment was reviewed by the author who coded the actual student solution to indicate points that the expert identified as a point where their thinking was flawed or the understanding was incomplete. For each assignment, the student and expert analyses were compared. This analysis is critical because it indicates whether students can self-generate appropriately targeted feedback on their homework solutions. The agreement was tracked for the entire semester to see whether agreement changes as the students become more familiar and comfortable with the wrappers. These results may also suggest ways to modify the wrappers to improve agreement.

The data were also analyzed to track individual students over time. The number of times a student made the same conceptual mistake in successive assignments or exams was recorded. The results for students who had self-identified that mistake were compared to results for students who failed to do so. Emergence of mistakes (following a previous assignment where that mistake was not made) and recurrence of mistakes were also classified as either problem-type identification or problem formulation procedure. It was expected that problem-type identification was more likely to emerge or recur, especially on exams that were longer after the homework assignments, because it is more difficult to provide authentic practice in problem-type identification.

594817 Process Design Versus Product Design: A Comparison of Student Problem-Solving Skills across Capstone Design Courses

Abstract:

Problem-solving is consistently cited as the most important technical skill that engineers learn during their undergraduate education. However, research suggests that students may not be learning how to solve the challenging and complex problems they will encounter in the workplace. Research on solving such "authentic" problems has been limited by: (1) the lack of an empirical framework for how expert engineers solve authentic problems and (2) the lack of any assessments that measure authentic problem-solving. In previous work, we developed an empirical framework of authentic problem-solving through semi-structured interviews with experts scientists and engineers from a variety of technical disciplines. This framework represents authentic problem solving as a set of decisions that experts make, guided by a cognitive construct we call a "predictive framework," which allows experts to explain observations, make predictions, and conduct mental simulations. We then developed an assessment in the context of chemical process design that requires participants to make some of these expert decisions and use these predictive frameworks, thus representing a measure of authentic problem solving. We use this assessment to quantitatively measure differences in problem-solving outcomes between two different capstone design courses at a large, private research university – one in product design, and one in plant design. Preliminary results suggest that students in the plant design course produce superior process design schematics, but that students in both courses are equally poor at identifying important features of the problem, and evaluating information that is relevant to solving the problem.

602350 Electric-Field Assisted Modulation of Surface Thermochemistry

Abstract:

Conventional catalyst design has enhanced reactivity and product selectivity through control of surface thermochemistry of reaction intermediates by tunable surface composition and surrounding environment (e.g., pore structure). In this work, the prospect of electric field towards controlling product selectivity and reaction networks on the Pt (111) surface was evaluated with periodic density functional theory (DFT) calculations in concert with machine learning algorithms to predict adsorption energies. Linear scaling relationships (LSRs) were developed for adsorption energies of surface species common to organic transformations on a Pt (111) surface. These electric-field linear scaling relationships were shown to be: (i) distinct as compared to zero-field relationships across metals for the same adsorbates, and (ii) linearly correlate with H* adsorption energies on Pt rather than the adsorption energy of the binding element. The LSRs for scaling slopes (i.e. slope of the LSR with adsorption energy of H* at atop site) or work-function with zero-field dipole moment enabled the estimation of field-dependent adsorption energy with the aid of field independent parameters (i.e. zero-field dipole moment of adsorbate and H* at the atop site or work-function). A random forest machine learning regression algorithm predicted DFT-computed adsorption energies with a mean absolute error (0.12 eV) comparable to the error of DFT. Overall, this study identifies the path forward for experimental and computational exploration for utilizing electric fields for catalysis, specifically towards effective catalyst stability, product selectivity, and control of reaction pathways.

594725 Effect of Potential and Explicit Aqueous Media on Ammonia Oxidation on Pt(111) Using Computational Multi-Scale Modeling.

Abstract:

Fossil fuels as a conventional source of energy have caused many environmental problems. Hydrogen is a clean and renewable source of energy; however, challenges regarding storage and transportation make it infeasible. Alternatively, water and ammonia contain significant amounts of hydrogen; further, they bypass challenges in transportation and storage since they can easily be converted to liquids. Ammonia oxidation requires only about 5% of the cell voltage (0.06 V) of water oxidation to produce hydrogen, which makes it more viable. It is also a promising way to denitrify wastewater. Additionally, ammonia can be used in a fuel cell to produce energy directly. However, catalyst costs and poisoning impede the broader adoption of NH_3 as a fuel. In electrochemical systems, there is another challenge; there is no electroneutrality due to presence of bias in the system. Solving these problems requires optimizing electrodes, electrolytes, and operating conditions. Molecular simulations play an integral role in this regard by providing molecular-level understanding of the electrocatalytic thermodynamics and kinetics. In this work, we use multiscale modeling to interrogate how system bias influences the free energies of electrocatalytic species. Specifically, we use a combination of density functional theory (DFT) and classical molecular dynamics (MD) to calculate the enthalpies, entropies, and free energies of species on a platinum electrode surface in the presence of both solvent and potential. The potential in the system is simulated using an explicit method capable of exchange of ions and electrons. We demonstrate the influence that electrode potential and pH have on the free energy of species in the NH_3 dehydrogenation pathway and compare these results with the analogous systems under thermal catalytic conditions. Through this exercise, we demonstrate creation of electrical double layer and the intense electric field in the vicinity of the surface and their influence on the free energies of catalytic species.

611410 Lithium Sulfur Batteries and Graphene Based Materials

Abstract:

My research background is a mixture of versatile disciplines including materials and chemical engineering, biomaterials and tissue engineering. My current research focuses on layer-on-layer cathode materials and gel electrolyte in ***lithium sulfur batteries*** in one set and on ***graphene-based materials*** in another set of projects. Towards DOE EV target, I have worked on preparation of high loading sulfur cathodes to increase the energy density, and gel electrolyte to improve battery safety. Moreover, I fabricated pouch cells in addition to coin cells to study the industrial challenges for our Li-S batteries. Graphene fibers, constitute another part of my research. I have developed a simple and entirely water-based yet effective method to fabricate graphene fibers with excellent electrical, thermal and mechanical properties. The ultralight GFs were deployed in epoxy composites rendering them to float in the water unlike the GF-free composite.

Prior joining the Robert Frederick Smith school of Chemical Engineering at Cornell University I worked on ***high internal phase biopolymer/protein emulsions*** and their application in food industry. I also investigated **protein/surfactant/biopolymer interaction** in the Department of Food Science at Cornell University. During this time, I had the opportunity to visit a materials science lab in the Department of Materials Science and Engineering to work on living radical polymerization and emulsion polymerization.

Biodegradable polymers and elastomers were the focus of my research during my PhD and my first postdoctoral fellowship. As a postdoctoral fellow at Qatar University working in collaboration with University of Bristol my research was on ***photocurable biodegradable elastomer*** synthesis for protein delivery, UV-crosslinking, and their thermal, mechanical and chemical characterization. Further, I opened a new route for research in that lab and defined projects based on ***electrospun fibers*** for Master's and undergraduate students. Finally, my PhD thesis was to synthesis ***biodegradable triblock copolymers*** and investigate the core-shell interactions of pH-sensitive block copolymer nanoparticles for drug-delivery.

Selected Publications and Presentations:

- L. Shebert[+], **S. Zamani[+] ([+]co-first authors)** C. Yi, Y.L. Joo, End-of-discharge failure and recovery of Li-S batteries with liquid and gel electrolyte: modeling and experiments (J. Materials Chemistry A, **2020**, 8, 4341).
- AlAmer, **S. Zamani**, K. Fok, A. Satish, Y.L. Joo, Facile Production of graphenic microsheets and their folding via water-based, surfactant-aided mechanical deformations (ACS Applied Materials & Interfaces, **2020**, 12, 8944).
- L. Joo, C. W. Chang, **S. Zamani**, 2019, Method for storage or transportation of graphene oxide, ***US patent*** Application No. 16/381,519, April 11, **2019**, Pending.
- Zamani[+], **S. Zamani[+] ([+]co-first authors)**, Z. Zhang, A. Abbaspourrad, Exceptional colloidal stability of acidified whey protein beverages stabilized by soybean soluble polysaccharide (*Journal of Food Science*, **2020**, 85, 989).
- **Zamani**, G.L. Shebert, C. Yi, Y.L. Joo, Investigation of gel-electrolyte failure mechanism in lithium-sulfur batteries (Meeting abstract, AIChE Annual Meeting, November 10-14, **2019**, Orlando)

- **Zamani**, G. Shebert, R. Zhang, Y. Joo, "Effect of Gel electrolyte on polysulfide transport in Li-S batteries" (Meeting Abstract, The 235th ECS meeting, Dallas, TX, May 2019)
- **Zamani**, H. Ismaeil, M. Elrayess, W. Kafienah, H. Younes, "A Biodegradable Photo-Crosslinked Fibrous Elastomer for Cardiac Tissue Engineering Applications" (Meeting Abstract, International Conference on Biopolymers and Bioplastics, Baltimore, MA, March 2019).
- Hassouna, **S. Zamani**, H. Younes, "Synthesis & Characterization of Tricarballylate Based Thermally crosslinked Elastomers for Drug Delivery & Tissue Engineering Applications" (Materials Science and Engineering C, **2018**, 93, 254-264).
- **Zamani**, N.M. Malchion, M.J. Selig, A. Abbaspourrad, "Formation of shelf stable Pickering high internal phase emulsions (HIPE) through inclusion of whey protein microgels" (Food & Function, **2018)**.

Under submission/review:

- **Zamani**, J.S. Won, M. Salim, M. AlAmer, C.W. Chang, A. Satish, P. Kumar, K. Amponsa, Y.L. Joo, Ultralight-weight graphene fibers with superior properties via entirely water-based process and their self-floating composites (under submission to Advanced Functional Materials)
- **Zamani**, C. Yi, G.L. Shebert, and Y.L. Joo, High sulfur loading cathodes with a layer-on-layer structure in Lithium-Sulfur (Under preparation to be submitted to Nano Energy)
- W. Chang^{+}, **S. Zamani**$^{+}$ **($^{+}$co-first authors)**, H. Suh, S. Park, T. Kang, Y.L. Joo, Cost-Effective Handling and Transportation of Graphene Oxides: Folding and Redispersion of Graphene Oxide Materials (under submission to ACS Nano)
- AlAmer, M. Salim, **S. Zamani**, C. Klassen, K. Fok, A. Umasankaran, Y.L. Joo, Curvature Effects on Graphite Exfoliation in Outer-Roatating Taylor-Couette Reactor

610174 Electrochemical Production of Ammonia Via Designing New Catalysts and Processes

Abstract:

Meeting our future global energy needs in an environmentally responsible way is one of the most significant challenges in the twenty-first century. Currently, fossil fuels such as oil, gas, and coal satisfy more than 80% of the global energy demand. This results in more than 35 billion metric tons of carbon dioxide (CO_2) emission annually. It is imperative to harvest renewable energy sources (e.g., solar, wind) to serve as a way to diversify from traditional fossil fuels for combatting the environmental challenges associated with greenhouse gases. Due to the intermittent nature of renewable energy sources, the development of a cost-effective and sustainable method of storing this vast amount of energy on an industrial scale when supply exceeds demand in the grid is an urgent need. As the cost of renewable electricity continues to decrease given the rapid progress in technology and economies of scale, there is a growing interest in fuels and chemicals electrosynthesis. This talk focuses on developing novel electrochemical systems and catalysts to use renewable electricity as an energetic driving force to convert low energy molecules (N_2) to high value-added molecules (NH_3) that can be utilized as either fuels, energy storage molecules, and/or chemicals.

598182 High-Fidelity Electrochemical Battery Models: Review, Issue, and Application

Abstract:

Today, lithium-ion batteries are considered the best means to store energy in renewable energy applications. When lithium-ion battery systems are deployed in energy applications, mathematical battery models are used to understand, operate, and optimize battery performance. Therefore, it is essential to use adequate and relevant battery models to improve cost, performance, and usability of energy applications. Typically, battery models can be classified into various groups, from empirical to molecular-level models. Among many other models, physics-based battery models hold the potential to simultaneously provide systems engineers, battery designers, and material scientists with an optimal computational framework that maximizes safety and usability for renewable energy applications and sustainable mobility.

In this presentation, physics-based battery models, related issues, and its applications will be presented. First, a pseudo-two-dimensional (P2D) and its simplified models will be reviewed. The P2D model is the most popular and acceptable physics-based model, and many simplified models have been proposed. Next, mathematical and numerical issues with physics-based models will be studied. Lastly, an efficient way to integrate physics-based models into renewable grid applications will be introduced. Typically, oversimplified battery models have been adopted for renewable energy applications, leading to system abuse and premature end of life estimate. Physics-based models can maximize the usability and performance of battery systems as well as renewable grid applications.

603717 Electrospinning for the Fabrication of Electrochemical Sensors

Abstract:

The use of exclusive nanostructure has gotten considerable importance in the biosensing and biomedical research. Lately, the electrospinning has drawn great attention in the development of nanofiber-based biosensor platforms. Electrospinning has emerged as an efficient nanofabrication technique for manufacturing of polymer, metal and metal-oxide nanofibers. Electrospinning emerges to be the ultimate technique to generate biocompatible nanofibers, such as biopolymers, carbon, ZnO, TiO_2 and NiO, etc. for the use in highly sensitive biosensing. Biosensors with enhanced sensitivity are becoming appealing for the sensing of biomolecules, and in particular, for glucose, cholesterol, triglyceride, and low-density lipoprotein sensing. Electrospinning is capable of depositing three-dimensional porous nanofibrous network on the surface of transducers in the sensor platforms, which offers high surface area, large global pore volume, predictable pore size distribution, and tunable interconnected porosity. These characters of nanofibers can add further desired functionalities to a sensor for the detection of bio-analytes in a specific environment, where efficient mass transport is necessary towards the electrode surface. Based on the distinctive properties of electrospun nanofibers in the nanoscale that differentiate them from other nanostructure created by other existing methods, we describe here the knowledge on nanofibers suitable for electrochemical biosensing applications, including structure and property characterization. Moreover, information on polymers together with carbons, metal and metal-oxides precursors and their processing conditions for the electrospinning of metal-oxide nanofibers is briefly described. Additional relevant issues regarding the research challenges, technology limitations, and future trends are also deliberated.

603560 Probing the Lithium Metal Anode Surface with First Principles

Abstract:

Batteries which employ a metallic Li anode are among the most promising technologies for next generation battery chemistries, owing to their high theoretical energy capacities (3860 mAh/g vs. 370 mAh/g for graphite anodes in state-of-the-art Li-ion technologies). Nonetheless, the implementation of Li metal anodes has been stifled by several key factors, many of which can be attributed to a lack of understanding of the surface phenomena occurring at the anode. More specifically, the native oxide layer is predicted to be present in any battery employing a Li metal anode. Yet, little is known about the structure of the Li metal/native oxide interface and its effect on the performance of Li metal batteries. In this work, first-principles calculations were used to characterize the atomic and electronic structure of the Li metal/native oxide interface.

Multiple models were generated for the Li/lithium oxide (Li_2O) interface, including crystalline and amorphous structures. With regard to the crystalline structures, the oxygen-terminated Li_2O interface yielded a lower interfacial energy and exhibited a work of adhesion more than 30 times larger than the lithium-terminated structure. These results suggest that the oxygen-terminated Li_2O surface is a more probable model under equilibrium conditions and that Li will more evenly coat an oxygen-terminated Li_2O surface during plating. In addition, a more realistic model of the interface, the amorphous Li/Li_2O model, was created using an *ab initio* molecular dynamics (AIMD) scheme to sequentially add O_2 molecules to a Li metal surface (mimicking the oxidation process in nature). The electronic structure of this amorphous interface illuminated behaviors that the crystalline interface failed to predict. In particular, considering atomic charges and shifts in Li 1s core-level electron binding energies, the size of the interfacial region is larger than that determined by a visual depiction of the atomic structure alone. Finally, the transport of Li ions through the amorphous interface were quantified with AIMD. To our knowledge, this work represents one of the first detailed analyses of the native oxide layer. These findings provide a set of fundamental limitations on batteries that employ high-capacity anodes based on lithium metal.

598069 Electrochemical Biosensors for *In Vivo* Neurochemical Analysis

Abstract:

The key to understanding the underlying mechanism of complex behaviors and disorders at the molecular-level relies on the ability to monitor neurotransmitter release and metabolite levels in the brains of freely moving animals. Therefore, the capability to simultaneously monitor multiple analytes *in vivo* would greatly facilitate our understanding of the connection between neurochemistry and behaviors. Previously reported analytical methods, such as microdialysis and fast-scan cyclic voltammetry, either did not have adequate temporal/spatial resolution or require specialized instrumentation.

To this end, we designed and fabricated a permselective membrane-based amperometric biosensing platform, targeting a group of non-electroactive analytes (*e.g.*, glucose, glutamate, and choline), which showed high sensitivity, selectivity and fast response time (*e.g.*, less than a few seconds).[1-4] The developed sensing platforms were successfully transferred to silicon-based microelectrode array microprobes. The underlying sensing mechanism was based on the electrochemical quantification of hydrogen peroxide, which is the end-product of oxidase-based enzymatic reactions. We achieved sensor selectivity against interferents by modifying the electrode surfaces with suitable permselective polymers (*e.g.*, polyphenylenediamine and Nafion). The deposited polymers acted as size and charge exclusion membranes, which allow hydrogen peroxide molecules to pass through to the electrode surface while excluding common interferents (*e.g.*, ascorbic acid and dopamine) present in the brain extracellular fluid. Additionally, we created an electrochemically deposited and nano-based on-probe iridium oxide (IrOx) reference electrode, eliminating the external bulky and inflammatory Ag/AgCl reference electrode.[1-3]

Leveraging the developed biosensors, on-probe IrOx reference electrode, novel PDMS-based thin-film transfer and enzyme microstamping technologies, we provided a unique approach to fabricate complete multifunctional neural probes.[1] The microprobes were able to simultaneously detect multiple analytes *in vivo* and deliver agents *in situ* and in a controlled manner. The integration of essential functions on a single platform removes the need for implanting multiple probes, with the dual benefits of reducing brain damage and surgical complexity. Furthermore, to minimize the mechanical mismatch between soft neural tissues and implants, we presented a liquid metal-enabled flexible and stretchable neural probe with ultra-large tunable stiffness for deep-brain electrochemical sensing and agent delivery.[2] Our flexible microprobes displayed mitigated inflammatory responses *in vivo*, as compared to silicon-based devices.

Reference

[1] B. Wang*, X. Wen*, *et al.* “A Multifunctional Neural Microprobe for Simultaneous Multi-Analyte Sensing and Chemical Delivery,” *Lab on a Chip*, **2020**, 20, 1390-1397.

[2] X. Wen*, B. Wang*, *et al.* “Flexible, Multifunctional Neural Probe with Liquid Metal Enabled, Ultra-Large Tunable Stiffness for Deep-Brain Chemical Sensing and Agent Delivery,” *Biosensors and Bioelectronics*, **2019**, 131, 37-45.

[3] B. Wang*, L. Feng* *et al.* “A Complete Electroenzymatic Choline Microprobe Based on Nanostructured Platinum Microelectrodes and an IrOx On-probe Reference Electrode,” *Electroanalysis*, **2019**, 31, 1249-1253.

[4] B. Wang*, B. Koo* *et al.* “Microbiosensor Fabrication by Polydimethylsiloxane Stamping for Combined Sensing of Glucose and Choline,” *Analyst*, **2018**, 143, 5008-5013.

*These authors contribute equally.

595927 An Educational Bioprocess Simulator (BioVL): Introduction of Disturbances

Abstract:

Simulators and virtual environments are increasingly used for educational purposes. However, commercial simulators – such as AspenPlus, HYSYS or Pro/II - have several issues in their design and application when deployed in the classroom. One of the key issues for undergraduate students is the sense of loss of reality with regards to the data[1]. This is due to the disparity between the data obtained in the laboratory, which comes with noise and errors; and the results from the solution of a mathematical model inside a computer-aided tool. The lack of variation and reality in the inputs and outputs of the model decreases the motivation of the students[2] and thus, it plays an important role in the user experience for these simulators as educational tools.

In the case of bioprocesses, disturbances do not only come from errors and noise, but also from the fact that we are dealing with living cells, and accordingly, there is intrinsic variation. Therefore, undergraduate students must be aware of the presence of variations inside this kind of processes and educational simulators must represent this reality. In this work, disturbances are implemented in the mechanistic models describing bioprocesses inside the educational software platform (BioVL) developed at the Department of Chemical and Biochemical Engineering of the Technical University of Denmark. BioVL[3] has been designed by integrating: 1) a thoughtful learning design; 2) a motivational approach based on the use of gamification; and, 3) the use of object-oriented modeling together with mechanistic models. The above-mentioned disturbances are based on superposing sinusoidal oscillating perturbations[4] inside the models. Through the linear combination of sine and cosine waves, we are able to introduce generic and time dependent disturbances inside the model. The function implemented produces realistic looking noise that is similar to industrial observations, based on an expert opinion. In the BioVL, the presence, amplitude and frequency of the disturbances are decided by the students and the noise can be removed from the results by clicking in the generated plot.

This open-source software has been developed in Python, where future users - the students - are involved as co-designers. The students have access to some of the mechanistic models and the disturbance function, which are available and explained in a Github repository (https://github.com/simonetacannodelas/BioVL-Library). BioVL is still a prototype and a user experience (UX) is scheduled for the beginning of May, where we will distribute it amongst a pool of students as beta users and co-designers. The results from the survey about the student's perceptions will be presented and addressed. Furthermore, it is the expectation of the authors that the addition of disturbances inside the models is not only valuable from a pedagogical perspective but also for the training of engineers inside production facilities. Hence, this software tool targets the education of undergraduate students in Chemical and Biochemical Engineering; although, it could be used as well for the training of operators and engineers in production in the future.

(1) Balamuralithara, B.; Woods, P. C. *Virtual Laboratories in Engineering Education: The Simulation Lab and Remote Lab.*Comput. Appl. Eng. Educ. **2009**. https://doi.org/10.1002/cae.20186.

(2) Feisel, L. D.; Rosa, A. J. *The Role of the Laboratory in Undergraduate Engineering Education*. J. Eng. Educ. **2005**, *94* (1), 121–130. https://doi.org/10.1002/j.2168-9830.2005.tb00833.x.

(3) Caño de Las Heras, S; Gutschmann, B.; Gernaey, K. V.; Krühne, U.; Mansouri, S. S. *Facilitating Learning by Failure through a Pedagogical Model-Based Tool for Bioprocesses*; Elsevier Masson SAS, **2019**; Vol. 46. https://doi.org/10.1016/b978-0-12-818634-3.50305-2.

(4) Branson, J. *Time Dependent Perturbation Theory* https://quantummechanics.ucsd.edu/ph130a/130_notes/node410.html (accessed Feb 21, 2020).

605718 Persistence and Practice of Spreadsheet Skills Using an Interactive Textbook

Abstract:

Spreadsheets provide a framework for chemical engineers to effectively complete calculations and represent data graphically. Since spreadsheets have existed for decades, many products, e.g., Microsoft Excel and Google Sheets, are available across computing platforms. Many formulas, functions, and tasks are common across programs and platforms, so education related to spreadsheets can be generalized. Spreadsheets Essentials, published by zyBooks – a Wiley brand, is an interactive textbook that record student interactions. Reading is an interactive experience with animations, matching and multiple choice questions, as well as static definitions and tables. Across three cohorts, reading rates above 90% were observed when a small fraction of the final course grade was provided as an incentive. Students perform various spreadsheet functions and write formulas using over 120 auto-graded, randomized problems. Success on these auto-graded questions was discussed previously and will be expanded upon across multiple cohorts. In addition, a case study will be presented examining a new "copy sheet" button that was available for the 2020 cohort. The frequency of copy sheet clicks will be examined for basic spreadsheet skills as well as advanced topics, such as solver, interpolation, numerical integration, error, statistics, and matrix operations. Examining both attempts before correct and success across questions and cohort will quantify the utility of the new copy sheet feature.

599213 Machine Learning for Fluid Property Correlations: Classroom Examples with Matlab

Abstract:

Recent advances in computer hardware and algorithms are spawning an explosive growth in the use of computer-based systems aimed at analyzing and ultimately correlating large amounts of experimental and synthetic data. As these machine learning tools become more widespread, it is becoming imperative that scientists and researchers become familiar with them, both in terms of understanding the tools and the current limitations of artificial intelligence, and more importantly being able to critically separate the hype from the real potential. This contribution presents a classroom exercise aimed at first-year undergraduate chemical engineering students, where a task is set to produce a correlation to predict the normal boiling point of organic compounds from an unabridged data set of >6000 compounds. The exercise, which is fully documented in terms of the problem statement and the solution, guides the students to initially perform a linear correlation of the boiling point data with a plausible relevant variable (the molecular weight) and to further refine it using multivariate linear fitting employing a second descriptor (the acentric factor). Finally, the data are processed through an artificial neural network to eventually provide an engineering-quality correlation. The problem statements, data files for the development of the exercise, and solutions are discussed within the context of a MATLAB environment but are general in nature.

Reference: L. Joss and E. A. Müller, *J. Chem. Educ.*, 96(4) 697–703 (2019).

600570 Reinforcing Computational Thinking throughout Chemical Engineering Curriculum with Matlab and Simulink

Abstract:

Chemical engineering students learn various computational tools either in dedicated programming courses offered by chemical engineering departments and in mass courses offered by other departments. Some of the computational tools and programming languages used in chemical engineering programs are AspenPlus, ChemCAD, HYSYS, MATLAB, Python, C, C++, FORTRAN. MATLAB is a high-level technical computing language and Simulink is a graphical environment to design, simulate and test systems. MATLAB and Simulink are widely used in industry and academia. Industrial user stories/videos on how MATLAB is used in chemical engineering applications will be discussed as an instruction tool to motivate students learn programming and create a vision of what students can do with computational tools when they join the workforce.

Using computational tools frequently throughout to curriculum helps students retain their computational skills. This talk describes examples of how MATLAB and Simulink can be integrated to fluid dynamics, heat transfer, reaction kinetics, mass and energy balances, process control curriculum and how computation can be introduced in introductory chemical engineering courses with free, self-paced, online tutorials created by MathWorks.

Offering students hands-on, laboratory experience is crucial for chemical engineering education. Novel technologies such as "Internet of Things" help students gain hands-on experience by accessing data from remote devices. ThingSpeak is an Internet of Things (IoT) platform with MATLAB analytics. With ThingSpeak, programming, hardware and data analysis with real-world data can be introduced to chemical engineering laboratory courses as well as statistics/data analysis courses. With interactive machine and deep learning apps in MATLAB chemical engineering students can rapidly gain data science skills that are on high demand and instructors can apply flipped classroom techniques using self-paced, free Machine Learning and Deep Learning tutorials developed by MathWorks. Project-based learning can easily be introduce to chemical engineering curriculum with hardware projects based on ThingSpeak and MATLAB Mobile.

With growing classroom sizes, offering fast feedback to students become challenging in programming assignments. To address this challenge, MathWorks developed an automated grading tool, MATLAB Grader. Use of MATLAB Grader to assess chemical engineering programming problems will be demonstrated and tools to create interactive lecture content (MATLAB Live Editor) will be reviewed. A self-paced tutorial to train instructors and teaching assistants on teaching and learning tools created by MathWorks will be presented.

595958 Instructional Videos and Interactive Notebooks for Learning Coding Concepts in Chemical Engineering Analysis

Abstract:

In many departments, a numerical methods course is used to introduce students to the key mathematical concepts and computational methods underlying modern process simulation and optimization software. The variety of course materials and covered topics present two significant challenges. First, pedagogical practices in mathematics, such as using a simple example to illustrate a concept, are often at odds with our student's notion of an authentic engineering experience. Next, the coding proficiency of students entering this course may widely vary from having functionally no programming experience to a high level of competence regardless of established prerequisites. As a result, it may be quite easy for students to get lost in the details of resolving a simple bug and lack the confidence required to see, correct, and learn from a methodological mistake.

In this talk, we'll detail our efforts at the University of Connecticut to improve a numerical methods course using a two-fold approach which consists of deploying a series of notebook-based coding assignments and developing complementary instructional videos. Notebook-based assignments make use of literate programming environments in Matlab and Julia. These consist of a template containing a series of code fragments with expository text and images, along with explicitly defined educational outcomes focused around coding skills. These template assignments are scaffolded in a manner that allows the instructor to distribute the burden of re-acquainting students with basic programming concepts over a series of assignments without the need to reteach programming skills explicitly at the start of the semester. We have found that template assignments naturally make the instructor consider transparent educational outcomes with respect to programming skills. Rather than taking the more traditional immersion-based approach, we have aligned our scaffolded approach with modern pedagogical theories of programming instruction.

In addition to literate programming environments, the students are provided with a series of video modules which provide an in-depth tutorial. This allows the instructor to model how a complicated problem is abstracted to a simpler one and the appropriate subject-area knowledge can be applied to address it. Over the course of the semester, the scaffold is removed as students become more confident in their abilities and better able to address more open-ended assignments. This helps to alleviate the degree to which an instructor is constrained by substantially revisiting prior programming concepts. Preliminary results, based upon survey data and assessment, suggest that students opt in to this more structured approach, and when they do, they increase their competency and self-efficacy.

611664 Experience with Symbolic Algebra and Smart Computing in Reaction Engineering Courses

Abstract:

Experience in using symbolic algebra and smart numerical computations in chemical reaction engineering and bioprocess engineering courses using Mathcad and Matlab is discussed to improve understanding of process fundamentals. Emphasis is on obtaining solutions with increased confidence and higher precision and with minimum effort by understanding what type of problem is being presented. Students are encouraged to not introduce unnecessary complications when attempting to solve a problem and solve problems using optimal effort. Example problems considered for illustration deal with solution of linear algebraic equations (kinetic parameter estimation and properties of small and large reaction schemes and metabolic networks), nonlinear algebraic equations (equilibrium calculations for multiple reactions, isothermal and non-isothermal operations of well-mixed reactors with single and multiple reactions, and kinetic parameter estimation using nonlinear regression), ordinary differential equations (isothermal, adiabatic, and non-adiabatic batch and steady state continuous flow tubular reactors), and symbolic algebra (rate expressions for catalytic reactions and characteristics of reaction networks). Analysis of multi-reaction networks is based on properties of atomic and stoichiometric matrices. Atomic matrices provide information on the maximum number of and identity of independent reactions. For a given reaction network, a study of the stoichiometric matrix leads to estimation of the number of independent reactions in the given network, relations of the dependent reactions in the network to independent reactions, and understanding of the relations among rates of formation of species participating in the reaction network. A smart use of these relations obviates the need for solving conservation equations for all species in the reaction network, which typically are systems of nonlinear algebraic equations (steady state CSTR) and nonlinear ordinary and partial differential equations (steady state and transient operations of batch reactors and imperfectly mixed flow reactors). In case any independent reactions are missing in the reaction network, these can be identified from the atomic and stoichiometric matrices. Reduced order reactor models for large reaction networks can be obtained by applying pseudo-steady state hypothesis (PSSH) for highly reactive species, such as free radicals and various forms of catalytically active sites. This approach is also useful in flux balance analysis in networks of biochemical reactions in living cell cultures. Symbolic algebra is very useful in this effort. Mathcad is interactive, with user-friendly graphics capabilities. Students are encouraged to display results using graphics for improved learning and understanding. The capabilities of Mathcad are of significant benefit in accelerating the learning and strengthening the fundamental knowledge base.

611282 *Five* Semesters of Hands-on Unit Operations Laboratory Courses

Abstract:

The Chemical Engineering Pillars curriculum at the University of Pittsburgh, allows our student to take their core CHE classes in a 'block-schedule' which immerses them in four two-hour classes each week, with the fifth two-hour class each week spent in a hands-on unit operations laboratory course. The Pillars curriculum does not add more credits, compared to a traditional curriculum, but instead uses the hours more effectively through restructuring. Importantly, for the student, it provides larger blocks of time to actively engage in learning in the classroom with the support of the instructor – ***and*** allows a hands-on unit operations laboratory experience each semester for the students in parallel with their core courses.

This presentation will provide details of the structure of the laboratory structure (i.e. how to provide hands-on laboratory experiences for approximately 300-350 students each semester), an over-view of the laboratory experiments – how they support the core courses and how they build upon one another from sophomore through senior year.

607569 Developing Hands-on Skills: Building a Gas Delivery System

Abstract:

Many chemical processing plants contains thousands of feet of pipes and tubing, therefore it is very useful for chemical engineering students to gain hands-on experience working with a variety of pipe and tube fittings. A three-week long hands-on session was implemented during a Junior-level Chemical Engineering Laboratory I course to develop student skills with lab hardware. During the first lab session, students learned how to identify different fittings, tubing, and pipes, along with how to size and name each piece. In the following lab session, students learned how to use wrenches properly to connect different fittings, and students learned how to create new pieces of tubing with nut-ferrule assemblies. Finally, in the last session, students used the tubing that they created during the previous session to build a gas delivery system. This last session culminated in a laboratory experiment that included pressurizing a system using a gas cylinder, testing the system for leaks, and determining a way to collect data from the system to calculate the volume of a small reactor. As a result, students not only learned hands-on skills, but they also applied these skills to create a gas delivery system to collect data to perform thermodynamic calculations. Students completed this lab session with a better sense of the work that goes into building a fluid flow network, and in future labs that involve more complex flow networks, students are better able to identify the different components of these systems.

Moving From The Traditional Way To Hybrid Approach For Teaching Unit Labs

Paper # 604806

Rupak Dua
Hampton University, Hampton, VA, USA

Abstract:

The unit operations laboratory (lab) in chemical engineering offers a unique learning experience where seniors are exposed to real-world chemical processes. They get hands-on activity on laboratory scaled industrial equipment and better insight into chemical engineering fundamentals that they learned in their sophomore and junior years. In these labs, the students work in groups to conduct the experiment, collect and analyze data, and finally write and present their technical reports. Thus these labs helped toned students' ability to collaborate and improve their writing and oral skills. The problem-solving skills, a core competency skill for an engineering student, do not seem to be much appreciated through these unit labs. To rectify the situation and take full advantage of the unit lab in making the students' problem solvers, we taught the class in a hybrid approach mixing the traditional Model of classroom instruction with the flipped classroom model over a semester.

During a semester, students performed three-unit laboratory experiments. Two of them were taught in a traditional way: the instructor provided the background and standard operating procedures for the instrument of that particular experiment. Then, the students in groups conducted the experiment under the instructor's supervision. The third remaining unit laboratory was taught in a flipped classroom model. Each group gathers the background information about the particular laboratory they were assigned to and figured out the working of the instrument by themselves by going over the manual and concept from the book and other online resources in detail. For the first two labs, they were tested by the submission of the individual laboratory report at the end of the experiment. While for the third experiment, in addition to the complete group lab report, each group was assessed every week on building different sections of the report, including objectives and introduction, theory, materials, safety precautions, standard operating procedures, observations tables, results, discussion, and conclusion. Each group was provided with feedback on every section each week to ensure they are going in the right direction. Each group also makes a video, in the end, summarizing the working of their third laboratory before submitting their final group laboratory report.

Preliminary data show that the students are much more engaged and better understand the experiment that they performed using the flip-room class model. The traditional way of learning provided them a guideline on how to experiment in an organized manner. In contrast, the flip room class model provided them an opportunity to successfully conduct their experiment on their own without much instruction from the instructor. Thus, we recommend that both teaching methods tradition and flip classroom models improve the students' overall skills and learning outcomes. The proposed hybrid teaching approach is expected to make the students more adept in following a systematic approach towards identifying and solving a problem.

Introduction:

The Unit Operations Laboratories (Unit Labs) courses offered over a year in the Department of Chemical Engineering provide a unique learning experience where senior classification students are exposed to real-world chemical processes [1]. They get hands-on activity [2] on laboratory scaled industrial equipment and better insight into chemical engineering fundamentals that they learned in their sophomore and junior years. Thus, reinforcing the theoretical coursework through the medium of experimentations with laboratory sized apparatus [3, 4].

These laboratory courses are often taught traditionally. Usually, the instructor goes over the background information, covers the theory and the experimental details during a lecture in a class before starting any experiment. Then an instructor gives the demonstration of the investigation to be carried out. Finally, the students work in groups to conduct the experiment, collect and analyze data under the instructor's supervision, and write and present their technical reports. This traditional teaching model enhances students' ability to conduct appropriate experimentation, analyze and interpret data, and use engineering judgment to draw conclusions [5]. It also improves their writing and oral skills and teaches them how to work together safely and cooperatively, which serves another purpose for these unit labs. However, we observed a lack of students' ability to independently identify, formulate, and solve the engineering problem. They were not engaged as they were supposed to develop the experiment of their own.

To rectify the situation and make the class more engaging and take the unit lab's full advantage in making the students' problem solvers, the instructor taught the class in a hybrid approach. This paper describes the approach that the instructor took to teach the course in the hybrid manner, mixing the traditional Model of classroom instruction with the flipped classroom model over two semesters

Methods:

Three unit labs were taught per semester (Figure 1). Two were taught in the traditional way where the instructor provided the background and standard operating procedures for the instrument of that particular experiment, and then the students in groups conducted the experiment under the supervision of the instructor. In contrast, the third lab was taught in a learner-centered model where the students learn by themselves. Specifically, each group gathers the background information about the particular laboratory they were assigned and figured out the instrument's working by themselves by going over the manual and concept from the book and other online resources in detail.

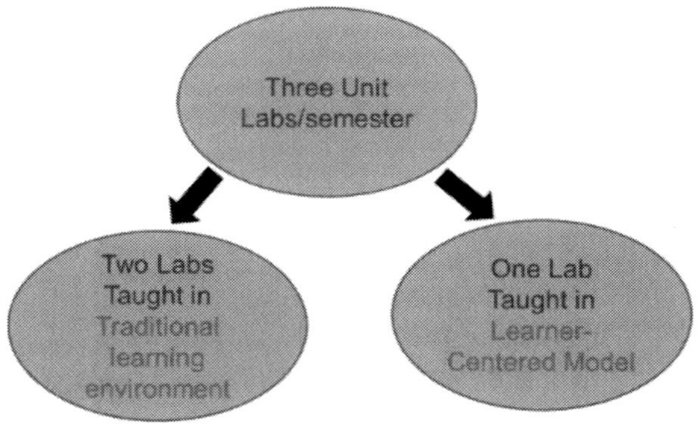

Figure 1. Hybrid Approach by teaching two unit laboratory experiments using Traditional Learning environment and one unit laboratory in

Assessment for 2 labs taught in the traditional way:

The assessment of each lab was spread over three weeks, first week, the instructor went over the lecture and demonstrated the experiment. The second week, the student worked in groups under the supervision of instructor. Finally, in the third week, they were supposed to submit their individual lab report by analyzing and interpreting their data.

Assessment for lab 3 taught in Learner-Centred Model:

The assessment of Lab 3 taught in the learner-centered Model was spread over the whole semester. In week 1, each group was assigned to a particular unit experiment and was given 3 weeks of time before anything was turned in for grading to the instructor. Different sections of the lab reports, including objectives, introduction, theory, material and methods, safety precaution, standard operating procedures (SOP), observation tables, results, discussions, and conclusions, were treated as individual assignments and were due at different weeks (Fig.2). At week 4, their first assignment (Objective and Introduction) was expected. Each group were given timely feedback on the part of their report, and that helped them to compile their complete group lab report due in week 14 and carried a weight percentage of 50% of the total lab 3 grade.

Lab 3 (Spread over the whole semester)

Week 1: Assignment of Unit Lab to the Group

Week 4: Objective and Introduction Due

Week 5: Theory, Material Methods and Safety Precautions Due

Week 6: Standard Operating Procedures Due

Week 9: Empty Observation Tables Due

Week 10: Results Due

Week 11: Discussion and Conclusions Due

Week 12: 5-minute video due

Week 14: Group Lab Report Due

Figure 2. Assessment for Unit Laboratory experiment 3 taught in Learnent-Centered Model

Post-course Survey:

At the end of the semester, a post-course survey was conducted in the class to evaluate the efficacy of this hybrid teaching approach. A questionnaire was prepared with 11 questions, 5 general questions about all 3 labs and 6 questions targeted for lab 3 taught in a Learner-Centred Model. The participants in this study were 14 undergraduate students of senior classification enrolled in the unit laboratory course at Hampton University. Student participation was not required by the instructor and was completely voluntary and anonymous.

Discussion and Conclusions:

The response rate got from the survey was 84.7%. We found that 81.8% of seniors taking this course liked working in groups and 72.7% of those found that the experiments taught in this course do not simulate the real-time environment. Further, we found that while 72.7 % of students like being trained in traditional teaching methods. However, more than half of the students responded (54.5%) that they learned more in the Learner-Centered Model while performing lab 3 because they were challenging. Everyone who responded encountered some problems while conducting the Lab 3 assigned experiment and agrees that's what happens in real life. Only 27.2% of students were able to run the Lab 3 experiment successfully. 100 % of the students who responded to the survey agreed that Lab 3 helped them understand what issues they might encounter when they go out in real life and help them think to find answers for problems. 63.6 % of students also agreed that if they have started working on their lab 3 sooner than later, they might run the experiments with more success.

Our survey data provided the preliminary results for the hybrid approach. We intended to conduct additional surveys from students in the Spring semester to have a long-term student thinking perspective and learning outcome. However, with the arrival of the COVID-19 pandemic in the middle of the Spring 2020 semester, we moved from an in-person unit lab teaching experience to a virtual environment to help save the community and the school. That disturbed our plans for getting another set of surveys.

Nevertheless, from the data we collected, we conclude that the traditional learning method provided students with a guideline on experimenting in an organized manner. Learner-Centered Model provided them an opportunity to successfully conduct their experiment independently without much instruction from the instructor. We recommend that both types of teaching methods are necessary to improve the overall skills of the students. Students are much more engaged and better understand the experiment that they performed using this hybrid approach. The proposed hybrid teaching approach is expected to make the students more adept in following a systematic approach to identifying and solving a problem and improving the learning outcomes.

References:

[1] B. Young, H. Yarranton, C. Bellehumeur, W. Svrcek, An experimental design approach to chemical engineering unit operations laboratories, Education for chemical engineers 1(1) (2006) 16-22.
[2] A. Hofstein, R. Mamlok-Naaman, The laboratory in science education: the state of the art, Chemistry education research and practice 8(2) (2007) 105-107.
[3] K. Russell, Bridging Theory and Practice on a Budget: A model for delivering practical knowledge through partnership with an on-campus facility.
[4] M.J. Zhang, C. Newton, J. Grove, M. Pritzker, M. Ioannidis, Design and assessment of a hybrid chemical engineering laboratory course with the incorporation of student-centred experiential learning, Education for Chemical Engineers 30 (2020) 1-8.
[5] A. Hofstein, M. Kipnis, I. Abrahams, How to learn in and from the chemistry laboratory, Teaching chemistry–A studybook, Brill Sense2013, pp. 153-182.

609018 Immobilization of Invertase in Calcium Alginate Gel Capsules and Determination of Batch Kinetics

Abstract:

Inverted sugar is an edible mixture of equal parts fructose and glucose typically produced by heating sucrose with water. This mixture is more desirable than sucrose because it does not crystallize due to its hygroscopic nature. The enzyme invertase, commonly isolated from *Saccharomyces cerevisiae*, is used to catalyze the hydrolysis reaction. Like many enzymes, invertase is time-consuming and expensive to isolate; therefore immobilization techniques have been developed to entrap the enzyme in porous gel capsules to make the process more efficient. Calcium alginate is a water-insoluble gel that can be safely made in the lab. This abstract outlines the procedure used to entrap invertase in calcium alginate beads. Additionally, once the invertase has been immobilized, a simple laboratory procedure to determine the kinetics of sucrose inversion in a batch system is outlined. The purpose of this abstract is to provide a methodology that can be used to create immobilized invertase gel capsules, and an analytical laboratory procedure that can be carried out by post-secondary students to produce calibration curves and determine the kinetics of invertase immobilization.

The work presented in this abstract outlines a junior/senior level chemical and biomolecular engineering laboratory module. In this module students immobilize invertase in calcium alginate gel beads using a technique that was outlined by Tanriseven and Dogan 2001 with a few modifications to ensure the enzyme will stay in the beads. The beads were refrigerated for one week and it was determine that their efficacy to convert sucrose to glucose and fructose was not affected over this period of time. The experiments were completed in 3 three-hour sessions (for a total of nine hours). In the first session, students created the calcium alginate gel beads. In the second session, a kinetics study was completed using batch reactors and commercial blood glucose meters to determine the amount of glucose present as a function of time in the reactors. And in the third session students filled a packed bed reactor with alginate beads and ran a simple experiment to determine the efficacy of the beads in a packed bed reactor system. Further work is currently being conducted to develop an analytical packed bed reactor model.

Using materials that are easy to find and relatively inexpensive, this lab introduces students to a variety of scientific concepts that can be explored to depths suitable for many different types of laboratory courses; development of a calibration curve, enzyme entrapment, batch reactions and enzyme kinetics can be taught with the laboratory experiment discussed in this abstract. This highly customizable experiment allows an instructor to introduce students to ideas ranging from basic scientific concepts like calibrating an instrument to big-picture industrial scale ideas of enzyme kinetics in packed bed reactors.

1. Tanriseven, A.; Dogan S. Immobilization of invertase within calcium alginate gel capsules. *Process Biochemistry*, **2001**, *36(11),* 1081-1083.
2. Hardee J. R.; Montgomery, T.M.; Jones, W.H. Chemistry and Flatulence: An Introductory Enzyme Experiment. *Journal of Chemical Education*, **2000**, *77*(4), 498-500.

605418 A Web-Based Unit Operations Lab for Enhancing Design, Control and Optimization in Chemical and Biochemical Engineering Courses

Abstract:

As we progress from delivery of instructional material from classroom to the internet, there is a need to develop virtual laboratory exercises to supplement the theoretical concepts covered in lectures. Many solutions are available on the Internet to deal with theoretical concepts remotely by the use of Learning Management Systems (LMS). However, solutions for laboratory exercises are usually scarce and most of the time they are substituted by theoretical exercises. Experimental exercises in a laboratory setting are really important in any discipline, but it is particularly relevant in engineering studies, where lab sessions are the connections between theory and practice by the use of real systems or semi-industrial processes. In the last decade, virtual labs have proven to be very versatile and valuable for providing practical experience for students in STE (Science, Technology and Engineering) fields (Alkhaldi, Pranata et al. 2016, Galan, Heradio et al. 2016, Heradio, de la Torre et al. 2016, Heradio, de la Torre et al. 2016, Potkonjak, Gardner et al. 2016). The main feature of these virtual labs is that they provide experiments that are based on realistic models of industrially important and technologically challenging process systems, what increases motivation of the students and at the same time provides relevant practical knowledge that enhances their educational experience.

There are many ways to provide web-based experimental experience. One approach is to create remote labs, which are web-based solution with access to actual experiments which can be operated remotely. Although this solution allows student to connect with the real system, it is limited to one user at a time which reduces its accessibility and it requires modifications in the real facilities to allow safe remote access and operation. A second approach is to develop virtual labs, which base the experiments on a virtual model of a system or process that mimics real-world behavior as much as possible. Such simulated exercises may be downloadable and run on a personal computer often using specialized software platforms such as Matlab or Aspen Plus. With the increased use of alternatives to personal computer such as tablets, smart phones and similar devices, the versatility of access and use can be improved if the virtual lab could be run entirely on the web without the need to download and use expensive software platforms.

In this presentation, we will discuss the development and use of a *purely web-based simulation lab* to teach concepts of design, operation, control and optimization of an anaerobic waste digester. The process is highly motivating to the students since it addresses two of the most pressing societal issues: waste biomass disposal and production of renewable energy. The anaerobic digester converts waste biomass into energetically useful biogas while producing organic manure.

The virtual simulation model of the Anaerobic Digester (AD) was developed using a flexible Javascript platform created at the University of Murcia in Spain called Easy JavaScript Simulations (Esquembre, 2015) and it based on well-known anaerobic digester models (Husain 1998, Haugen, Bakke et al. 2013, Haugen, Bakke et al. 2013, Haugen, Bakke et al. 2014, Xie, Hai et al. 2016). The objective was to supplement courses in process control and process design with a laboratory component. This virtual laboratory experience could be used on any device capable of accessing the web (laptops, tablets, smartphones etc.). It is designed to enable the student to explore the multiple facets associated with a complex biochemical system with a view to understanding its design, operation, control and optimization (see Figure 1). A user-friendly interface (see Figure 2) that allows easy specification and testing of operational parameters is provided. In this presentation, we discuss how the model and simulation is developed and deployed. We will demonstrate examples of its use in an educational environment.

References

Alkhaldi, T., I. Pranata and R. I. Athauda (2016). "A review of contemporary virtual and remote laboratory implementations: observations and findings." Journal of Computers in Education **3**(3): 329-351.

Esquembre, F. (2015) "Facilitating the Creation of Virtual and Remote Laboratories for Science and Engineering Education". IFAC PapersOnLine, **48**(29): 49–58.

Galan, D., R. Heradio, L. de la Torre, S. Dormido and F. Esquembre (2016). "Virtual Control Labs Experimentation: The Water Tank System." IFAC Papersonline **49**(6): 87-92.

Haugen, F., R. Bakke and B. Lie (2013). "Adapting Dynamic Mathematical Models to a Pilot Anaerobic Digestion Reactor." Modeling Identification and Control **34**(2): 35-54.

Haugen, F., R. Bakke and B. Lie (2013). "Temperature Control of a Pilot Anaerobic Digestion Reactor." Modeling Identification and Control **34**(3): 99-117.

Haugen, F., R. Bakke and B. Lie (2014). "State Estimation and Model-Based Control of a Pilot Anaerobic Digestion Reactor." Journal of Control Science and Engineering.

Heradio, R., L. de la Torre and S. Dormido (2016). "Virtual and remote labs in control education: A survey." Annual Reviews in Control **42**: 1-10.

Heradio, R., L. de la Torre, D. Galan, F. J. Cabrerizo, E. Herrera-Viedma and S. Dormido (2016). "Virtual and remote labs in education: A bibliometric analysis." Computers & Education **98**: 14-38.

Husain, A. (1998). "Mathematical models of the kinetics of anaerobic digestion - A selected review." Biomass & Bioenergy **14**(5-6): 561-571.

Potkonjak, V., M. Gardner, V. Callaghan, P. Mattila, C. Guetl, V. M. Petrovic and K. Jovanovic (2016). "Virtual laboratories for education in science, technology, and engineering: A review." Computers & Education **95**: 309-327.

Xie, S. H., F. I. Hai, X. M. Zhan, W. S. Guo, H. H. Ngo, W. E. Price and L. D. Nghiem (2016). "Anaerobic co-digestion: A critical review of mathematical modelling for performance optimization." Bioresource Technology **222**: 498-512.

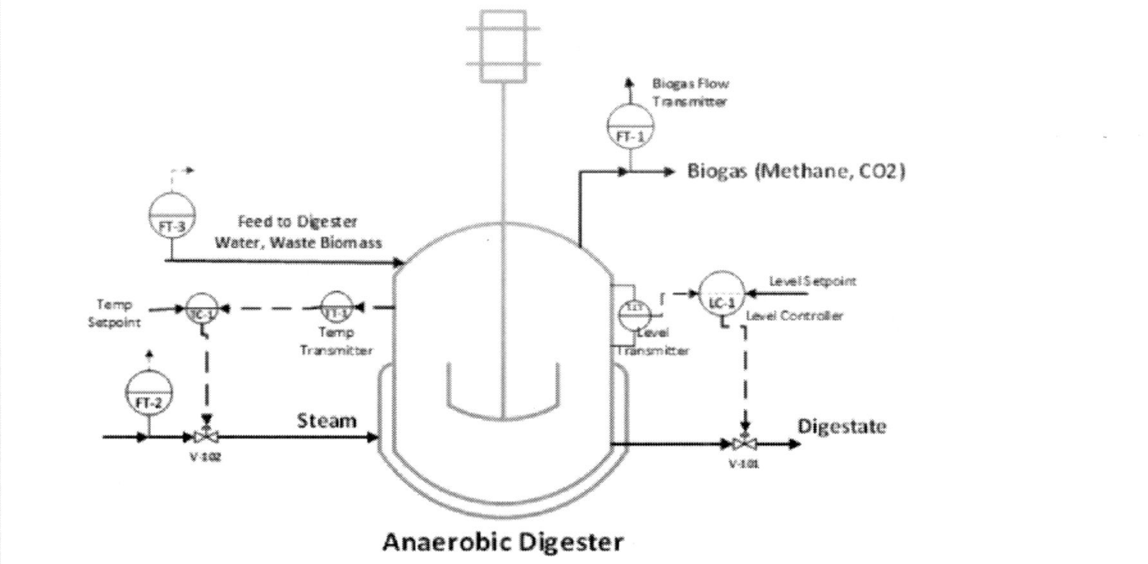

Figure 1. Control Instrumentation Diagram of the Digester.

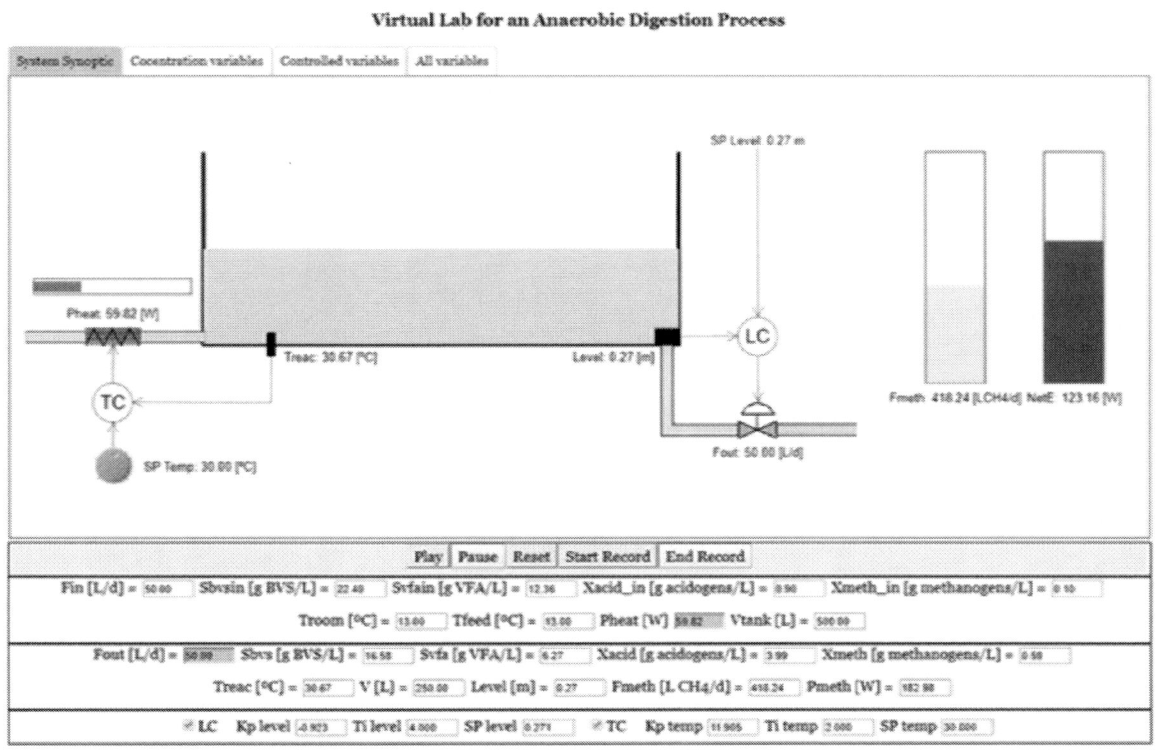

Figure 2. Virtual Lab Interface Web Page

605422 Meeting Unit Operations Laboratory Learning Outcomes in the Era of Covid-19

Abstract:

As chemical engineering students, unit operations (UO) courses represent an opportunity for students to gain hands-on experience with process-scale equipment in a laboratory. In addition to cementing the link between fundamental theory, safety and ethics, and industrially relevant operation, these courses allow students to see, hear, feel, and physically manipulate equipment and measure and observe the intended (and sometimes unintended) consequences of their actions. During the Spring of 2020, many faculty were faced with the challenge of giving students these same learning experiences through an online teaching format. While this quick change in instruction put many in "survival mode" to get through the remainder of the semester, we have used this experience as a learning opportunity to pool our collective knowledge on successful and unsuccessful practices for meeting the learning outcomes of a UO laboratory course online. In this presentation, we will share the knowledge that we gained from the Spring 2020 semester, with particular emphasis on effective online teaching and management strategies to meet learning outcomes for the UO lab inside and outside the formal laboratory environment. Additionally, we will focus on how we can engage students in active learning experiences related to process safety despite not being in a traditional lab setting. Overall, the Covid-19 pandemic was a learning experience for all on how best to continue to meet our learning objectives in a new online format. Moving forward, we should take the lessons that we learned on effective teaching practices in the online format and use it for improving curriculum in the classroom.

608348 Got Water? : A Filtration Experiment for the Material and Energy Balance Course

Abstract:

In Material and Energy Balances (MEB) students often learn about MEB but do not have opportunities to apply material balances concurrently with the class. With this hands-on project, students are presented with a simulated real-world problem and are challenged to create a water filter that purifies water contaminated with food coloring. Students are provided a 3-hour experimental session to test a variety of materials including charcoal of different grain sizes, steel wool and coffee filters to investigate their effects on creating clean water. A week after experimenting students return for a 30-minute demonstration session to produce the highest quantity and purity of water. Students are judged on the amount of water produced, the purity of the water and the cost of their filter.

601618 “Sustainabilty – the Need for Scalable Solutions & a Framework to Make Decisions”

Abstract:

Sustainability had taken a new dimension in recent times with employees and investors challenging industries to be more sustainable. An overview of CO2 generation sources globally and the USA will be presented and a case will be made for the need to scaleable solutions. The scalability and mutual interaction of these solutions have resulted in trade-offs. In this talk, the need for a frame-work for decision making and the role scientists and engineers can play is discussed.

606476 The Impact of Policy Instruments on Deployment of Low-Carbon Heating in the UK: A Whole System Optimisation Study

Abstract:

As part of the UK's commitment to emitting net zero greenhouse gases by the year 2050, the country must decarbonise the space heating sector, which currently contributes 17% of these emissions. This will involve retrofitting 24 million residential buildings, currently using natural gas, to a low carbon form of heating. The main options to achieve this are either hydrogen based, such as hydrogen boilers, or electricity based, such as electric heat pumps and electric resistive heating. Each option involves considerable costs along the supply chain, including upgrading infrastructure and installing new equipment in homes.

One of the key challenges facing policymakers is that, at present, natural gas heating is often the cheapest form available to households. In addition, considerable investment will be required in hydrogen production and transportation before it becomes available to consumers. Consequently, the transition away from natural gas heating is likely to require financial incentives. Each of the strategies is likely to present different costs to different stakeholders and so policies will be required targeting different points of the supply chain to incentivise the uptake of low-carbon heating.

The aim of this study is to assess how policy instruments, designed to promote the uptake of renewable heating, might impact both the heating technology mix and the supporting infrastructure when considering the least cost solution to decarbonisation. The Value Web Model, a mixed integer linear programming model for integrated energy value chains, was developed to optimise the deployment of heating technologies alongside energy generation, transportation, and storage to meet energy demands in the UK. This model considers natural resource availability (including wind, solar, land, and natural gas) alongside energy demand with high spatial and temporal resolution to determine not only which technologies should be used, but where they should be located. The results will show how different incentives applied at different points along the supply chain and in different amounts impact both the penetration of the technology they are aimed at and the overall technology mix. This work will highlight the impacts of incentives on the wider energy system and add to the discussion, relevant to policymakers and researchers, on how to decarbonise energy systems at the lowest cost.

611169 The Role of Chemical Engineering in Sustainable Energy Policy

Abstract:

Technological advancements are often made long before their application is mandated by policy. Conversely, the way a technology is regulated can make or break its adoption. Sustainability of energy involves balancing our energy needs with the well-being of the environment and the community. This talk will examine examples of policy successes and how chemical engineering can contribute to sustainability of energy and energy policy.

600820 Review of US Environmental Protection Agency General Exposure Scenario Modeling for Evaluating Chemical Safety

Abstract:

Generic Scenarios were initially developed by the US Environmental Protection Agency (USEPA) as a part of the chemical risk assessments process required under the Toxic Substances Control Act (TSCA) used by the USEPA to evaluate the hazards associated with the modeled chemical. From these humble beginnings, Generic Scenario documents have evolved into Organisation for Economic Co-operation and Development (OECD) Emissions Scenario Documents (ESDs) produced to describe a chemical's sources, production processes, pathways and use patterns. This presentation provides a review of exposure modeling performed in the development of existing Generic Scenario documents. This review was conducted to understand the evolution of the Generic Scenario process, with a goal of simplifying the modeling approaches used to estimate exposure levels and human health effects related to manufacture and use of chemicals currently required under the Lautenberg Chemical Safety for the 21st Century Act.

597556 Effectively Communicating Sustainability and Engineering within Federal Policy

Abstract:

As engineers and scientists we wonder why it seems that science is void from the decisions made be policy makers and governmental entities. There are scientists and engineers in government, but we have to be willing to put in the work to be effective, and those implementing policy have to be willing to listen. The ideas presented in this talk will give insight into how varying levels can effectively communicate scientific and sustainable ideas to those in policy. Part of this talk will be describing the process for becoming a Science and Technology Policy Fellow of the American Association for the Advancement of Science, as well as the roles of a fellow in the Executive branch of the federal government and how sustainability can be communicated to an array of non-scientific audiences.

597865 Engineering Student Mental Health – Analysis of National Data from the Healthy Minds Study

Abstract:

Data from national studies indicate an increasing prevalence of mental health problems among college students. For instance, the frequency of depression increased from 24.8% in 2009 to 29.9% in 2017, while the percentage of students reporting suicidal ideation increased from 5.8% in 2007 to 10.8% in 2017 (Lipson, Lattie, and Eisenberg 2018). Similar increases were seen for anxiety from 22% in 2014 (Eisenberg et al. 2014) to 31% in 2019 (Eisenberg et al. 2019). While the problem is not unique to engineering students, engineering faculty have become increasingly motivated to better understand factors influencing the mental health of their students. In this study, publicly available data from the 2018-2019 Healthy Minds Study were analyzed to better understand how factors related to mental health differ between the engineering undergraduate population ($n \sim 4{,}000$) and the general undergraduate population ($n \sim 60{,}000$). From these data, 32% of engineering students were scored as having moderate to severe depression, as compared to 37% of the general student population. That being said, only 32% of undergraduate engineers with significant anxiety or depression symptoms had sought professional help in the last year, compared to 45% of their non-engineering peers, indicating a significant mental health treatment gap in engineering students. Factors that could influence student mental health or help-seeking behavior were also analyzed. For instance, engineering students spent more time in class, with 44% indicating more than 16 hours per week in class vs. 32% of non-engineering students. Similarly, 37% of engineering students indicated that they spent more than 16 hours per week on homework vs. 27% of non-engineering students. Data from this study could guide the design of research studies to identify targets for improving the mental health of engineering undergraduate students.

607713 Incorporating Diversity Equality & Inclusion Training into First-Year Chemical Engineering Curriculum

Abstract:

STEM fields, particularly engineering disciplines, have struggled to attract, welcome in, and retain students from a diversity of backgrounds. From the undergraduate population up to the faculty level, significant disparages still exist in engineering colleges along lines of race, ethnicity, gender, sexual orientation, and more. However, a department's climate for underrepresented groups can be difficult to change and negative behaviors or attitudes among students can be self-propagating; as such, setting the tone and expectations early for a student body is important.

We have integrated diversity equality and inclusion (DE&I) training into both semesters of our first-year curriculum, with the aim to set a first impression that communicates the importance of DE&I to our department and to our students' eventual careers as chemical engineers. One significant hurdle encountered early in these efforts was to maintain investment in these topics by students who are traditionally over represented in engineering fields. This work focuses on means to obtain early "buy-in" from *all* first-year students in our program and empower underrepresented students to address deficiencies in department culture.

In our fall semester Introduction to Chemical Engineering Course we use several strategies to promote DE&I for our first year students. Primary among these strategies is a three lecture series on engineering ethics; this totals about three hours of *interactive* material on engineering ethics, two-thirds of which focuses solely on DE&I as an integral part to engineering ethics. DE&I topics are introduced through the lenses and language of engineering to better connect to the target audience and emphasize their relevance.

In the second semester of the first year we have students in our Chemical Engineering Design & Innovation course, which is a much more active hands-on learning environment, in which students work in teams on open-ended team-based projects. It is in this course in which DE&I training is most often put to practice, because student will work on a team for about two weeks with almost every peer in their section. Significant efforts are made to establish a climate of inclusion in the makerspace and lab areas in which this course occurs, and students receive additional training in conflict resolution.

In this talk I will discuss the various DE&I methods implemented, which we have found to be effective in these two very different courses. I will also present some evidence of efficacy from student surveys, observations, and free-form responses. In sum, while there is always more DE&I ground to be gained, we see significant evidence that we have created a more welcoming, equitable, and inclusive environment for our first-year students, one that has reached into the later years of our curriculum.

602855 Comparing Student Performance and Preparation Indicators for a Summer Online Thermodynamics Course with a Full Term Face-to-Face Offering

Abstract:

Considering recent events, the ability to offer effective online courses takes on new importance. The authors have been teaching a core course in chemical engineering thermodynamics online for the past four summers, and have collected evidence which supports (albeit without statistical significance) the hypothesis that outcomes as measured by a common final exam are essentially indistinguishable from a regular term face-to face course offering. Perhaps most interestingly, the outcomes appear insensitive to the student's average cumulative GPA when entering the course.

This transitioned thermodynamics course coverage includes 1st and 2nd law (building off a pre-requisite material and energy balances course), equations of state, phase equilibrium, mixtures, and reaction equilibrium. This presentation will provide details on course structure, lessons learned for teaching foundational courses online, and data that supports other literature findings that a well-designed online course produces outcomes indistinguishable from face-to-face courses. The data extends these findings with indications that acceptable learning outcomes can be attained even when offered in a highly compressed summer term time frame.

598135 Hands-on in the Online Classroom: Transitioning a Hands-on Pedagogy to a Virtual Format

Abstract:

During the coronavirus outbreak in spring 2020, many universities across the United States chose to transition to online instruction for a significant portion of the semester. While the transition to online learning for traditional classes can be straightforward, determining how to best implement hands-on pedagogies in an online environment is substantially more challenging; however, these pedagogies can still be valuable for improving student conceptual understanding and motivation. Our team has previously developed a number of Low-Cost Desktop Learning Modules (LC-DLMs) for fluid mechanics and heat transfer courses which are typically utilized by small groups of students in a classroom setting. Students perform qualitative experiments, analyze data, and discuss conceptual items in an interactive format. This presentation will explore our efforts to transition this pedagogy to a virtual format. We created a detailed demonstration video showing data collection for each module, provided short instructional videos covering concepts defined in learning objectives, and transitioned worksheet materials to a format supported by online learning management systems. Results from pre- and post-implementation conceptual assessments and motivational surveys will be presented for virtual implementations at several universities for a hydraulic loss and double pipe heat exchanger LC-DLM. These results will be compared to conceptual and motivational data collected during traditional hands-on implementations of the same modules in fall 2019. We hypothesize that these results will show nearly equivalent effectiveness of virtual hands-on pedagogy implementations and we will highlight any significant differences in conceptual understanding or motivation compared to hands-on implementations. We also hypothesize that this early work will allow us to develop refined virtual implementations of hands-on pedagogies which will be broadly useful at resource-limited universities or in traditionally online-based programs in the future.

600687 Design Principles for Independent Experiential Learning Resources: A Review of Literature and Personal Experience

Abstract:

Introduction

Experiential learning exercises such as labs are almost universally acknowledged to be a core aspect of the training of any engineering student. It serves multiple purposes ranging from the demonstrations and validation of concepts taught in the classroom to enabling independent learning by students. A key part of the experiential learning process that is gradually getting more attention in pedagogical literature are the resources that the students use. These resources include the rigs and apparatus, various teaching handouts and online video resources that students can access. Resources intended for defined experiments with little student autonomy are conceptually more straightforward to design as they need to serve a singular function with limited variability. In contrast, independent learning resources are more complex as the resources need to help create an experimental design space for students to explore. These resources need to be carefully designed to incorporate features that facilitate the exploration and learning process.

The 2nd year teaching labs, named the "Knowledge Labs" at the Department of Chemical Engineering, Imperial College London has undergone a significant overhaul in 2016 in response to consistently poor student feedback. This process involved the redevelopment of the entire module and the in-house construction of 6 research-grade pilot scale teaching rigs and associated resources for topics ranging from reaction engineering to process control and separation processes amongst other modules. Following the overhaul, despite various teething issues, students have been consistently reporting significantly improved engagement and interest in the module. This experience, together with a review of recent literature serves as the basis of the identification and elucidation of the following design principles that we believe educators should incorporate in their own design of resources of independent experiential learning processes. Figure 1 outlines these design principles.

Principle 1: Flexibility

Two modes of flexibility in the lab can be facilitated by resource design. The first is **flexibility in design of experiments**. This mode means that students are empowered to conceive their workflow in a variety of ways and need to apply material from the classroom and engineering discernment to select and justify a workflow. The second mode of flexibility is **flexibility of concept integration**. This mode refers to a resource that can integrate various topics from different engineering domains (e.g. reaction engineering and process control) in the same workflow. Flexibility has been incorporated in the following examples:

- Providing students a highly flexible platform or a tool which then can be used by students in a wide variety of ways [1].
- Developing rigs that intrinsically draw upon different topics taught such as performing a chemical reaction in a rig that also needs control [2].

Principle 2: Accessibility

Accessibility is a broad design principle which can be understood as incorporating features and approaches that help students relate to and engage with the material. This principle is more effectively elucidated with examples:

- Adopting a participatory design approach where student input results in user-friendly features and student-accessible designs being incorporated [2].
- Using accessible contexts such as food and beverages (F&B) as the premise for the project [3].
- Providing extensive training in the form of a structured introductory example before tasking students with a more complex project [4].

Principle 3: Authenticity

Authentic learning activities can be classified into four categories: 1) context authenticity where the learning context is relevant to future professional practice, 2) task authenticity where the skills developed are relevant in professional practice, 3) impact authenticity where the task has an impact relatable beyond the classroom and 4) personal authenticity where the task resonates with the individual and has a personal impact [5]. Examples of authenticity include:

- Incorporating tools and features from professional practice such as industrial / research grade equipment, designs or tools which facilitates context and task authenticity [2].
- Using accessible and authentic contexts like F&B simultaneously facilitates all the modes of categories of authenticity [3].

Principle 4: Reliability

Reliability of the resources / project is essential for students to have a meaningful experience. When things do not work, students do not get the full learning experience and often stress excessively about their lab reports and grades. This was one of the main teething issues faced in the Knowledge Labs. Reliability can be achieved in the following ways:

- Beta test the resource to a level above and beyond what is expected of students so the resource is known to work at all reasonable operating conditions. The teaching assistant can also be involved in this process [2].
- Consider a virtual laboratory / simulations to either complement or replace physical labs as computational platforms by nature provide far more consistent results [6].

Principle 5: Safety

The safety of the resources and the overall project is an essential consideration as educators have an important obligation to students to provide them with a safe learning environment. This can be achieved in the following ways:

- Adopt professional practices in equipment design and activity risk assessments when design resources and the project [7]
- Involve students in performing risk assessments and use the opportunity to share how various safety features were included.

Analysis of Principles:

Kolb's experiential cycle is an excellent lens to use to understand how the various design principles when incorporated effectively activate various aspects of the learning cycle and therefore the entire learning process. Each of the principles are shown in Figure 2, with safety not directly impacting the learning cycle, but serving as an essential backdrop to which any learning should occur in.

Conclusion

The five design principles outlined here are the emergent product from the overhauling of the Knowledge labs and have been found to be the summation of the various efforts made to provide the best possible learning environment and experience to students. Efforts by other authors in designing experiential learning resources for independent learning have either consciously or unconsciously incorporated one or more of the five design principles. The consideration and inclusion of these design principles into the resource development workflow will aid the development of effective learning tools for students.

References

[1] E. Bogdanov, C. Salzmann, and D. Gillet, *Widget-based approach for remote control labs*, vol. 9, no. PART 1. IFAC, 2012.

[2] P. Inguva *et al.*, "Advancing experiential learning through participatory design," *Educ. Chem. Eng.*, vol. 25, pp. 16–21, Oct. 2018.

[3] M. Xie *et al.*, "Accelerating Students' Learning of Chromatography with an Experiential Module on Process Development and Scaleup," *J. Chem. Educ.*, 2020.

[4] V. V Acuna, R. M. Hopper, and R. J. Yoder, "Computer-Aided Drug Design for the Organic Chemistry Laboratory Using Accessible Molecular Modeling Tools," *J. Chem. Educ.*, p. acs.jchemed.9b00592, Feb. 2020.

[5] J. Strobel, J. Wang, N. R. Weber, and M. Dyehouse, "The role of authenticity in design-based learning environments: The case of engineering education," *Comput. Educ.*, vol. 64, pp. 143–152, 2013.

[6] B. Balakrishnan and P. C. Woods, "A comparative study on real lab and simulation lab in communication engineering from students' perspectives," *Eur. J. Eng. Educ.*, vol. 38, no. 2, pp. 159–171, 2013.

[7] M. Gunasekera, S. Ahmed, and F. Khan, "Integration of process safety in equipment design: A framework for academic learning activity," *Educ. Chem. Eng.*, vol. 30, pp. 32–39, 2020.

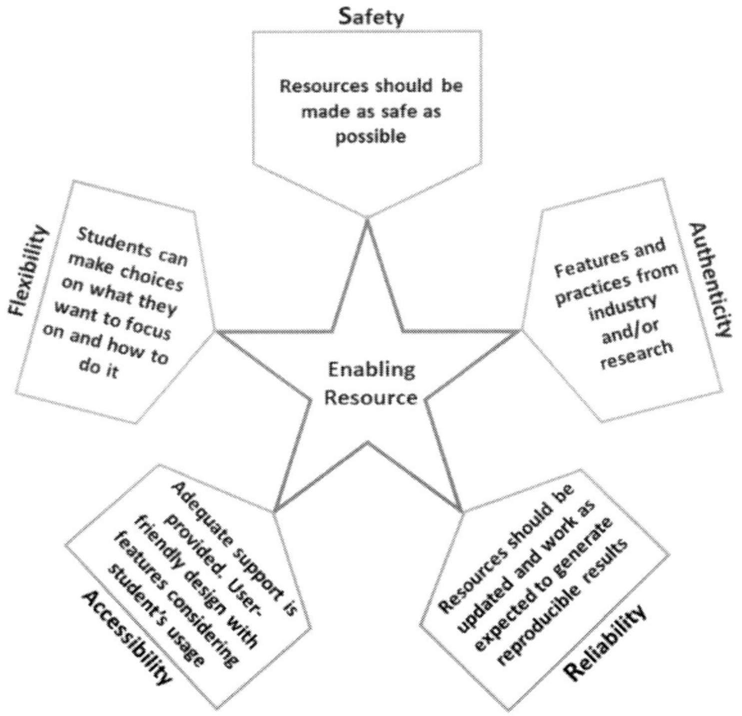

Figure 1. Schematic of the resource design principles.

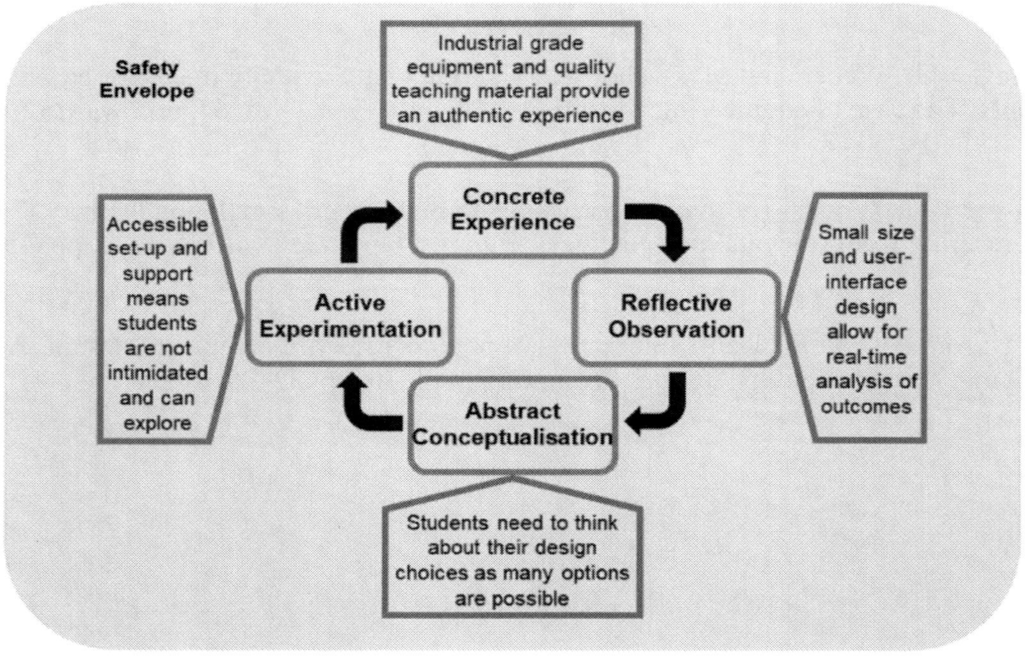

Figure 2. Impact of design principles on Kolb's experiential learning cycle.

599458 Modernizing the Chemical Engineering Curriculum Via a Student-Centered Framework That Promotes Technical, Soft, and Tech-Savviness Skills: The Case of Unit Operations

Abstract:

The development of a comprehensive set of skills is critical to succeeding in the increasingly competitive global job marketplace. An emerging workforce with technical, soft and tech-savviness skills is of the utmost importance to address some of the most relevant challenges in the years to come. In the Department of Chemical Engineering at Universidad de Los Andes, we decided to tackle such a challenge by pursuing a major curricular reform where a flipped-classroom approach, a PO-PBL problem, and interactive e-learning tools were developed and tested. The proposed intervention was first implemented in the core course of Unit Operations. Herein, we show that upon intervention, the students' perception regarding the development of teamwork and people-related skills were crucial to creating stimulating environments where students felt compelled to discuss and put concepts to practice. Additional analysis showed that diverging learning styles appear benefited by the proposed framework.

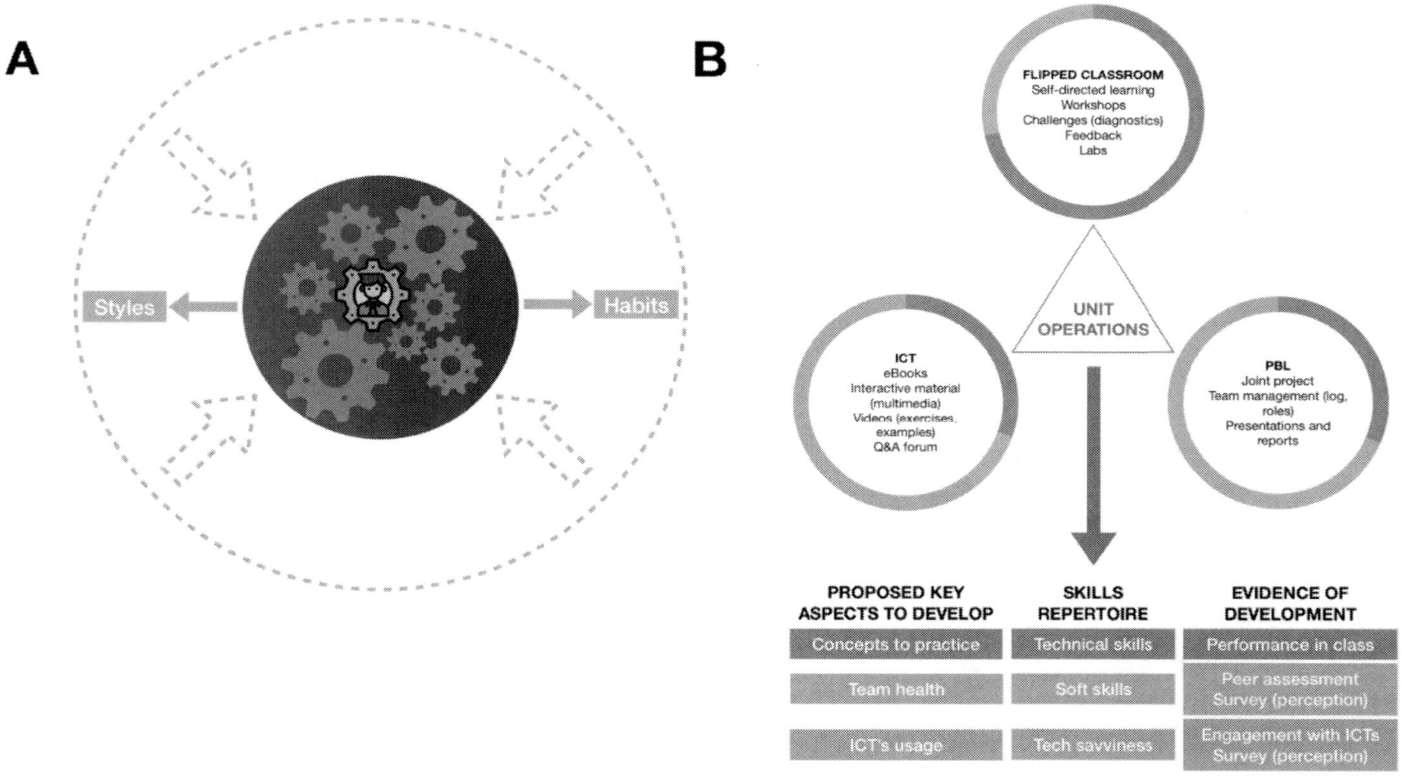

609997 Developing and Delivering an Undergraduate "Introduction to Research Methods" Course for Honors Students: Lessons Learned

Abstract:

Our institution has focused on expanding the Honor's College experience in an effort to improve student recruiting and retention. Within the College of Engineering and Applied Science (CEAS), the goal is to focus the Honors experience on undergraduate research. Undergraduate research is considered a High Impact Program (HIPs) for recruiting and retention. The aims of the CEAS Honor's program are to 1) broaden research opportunities, 2) improve competitiveness of student applications for summer research programs, NSF REUs, internal/external research funding applications, 3) increase participation in undergraduate research conferences, and 4) prepare the students for graduate school. Historically, many students have received credit for completing undergraduate research, but this is often a "stand-alone" course with no additional preparation and ill-defined outcomes. For this course, the research is strongly dependent on the research lab and research advisor. The new College Honor's program is also expected to lead to a significant increase in the number of students performing undergraduate research. Improving the efficiency and effectiveness of undergraduate research suggests a group mentoring approach. This will empower the students to take ownership of their research experience. In an effort to improve the undergraduate research experience and achieve the above outcomes, the CEAS initiated an "Introduction to Research Methods" course for the undergraduate Honors students as pre-requisite for undergraduate research during the Spring 2020 semester.

The goal of this course is to group mentor the students on many common research topics in advance of them entering the laboratory and performing research. This should reduce the required mentoring load of the research advisor and also lead to above desired outcomes. The Research Methods course will be broadly focused by providing a general approach to research and graduate school preparation appropriate for all majors in the Engineering College. Alternative approaches from the literature that are used to teach students how to conduct research will be compared and contrasted. Course topics will include: finding a research mentor, literature search skills, using the scientific method for approaching a research problem, developing a research methodology, research ethics, laboratory safety, writing a funding proposal, delivering a research presentation, and selecting and applying for graduate school. The motivation for this work, course details, learning objectives, course schedule, and course assignments will be presented. Assessments and student outcome tracking for the course will be discussed. Experiences, outcomes, feedback, and lessons learned from the initial offering of this course will be presented.

598232 To Pre-Req or Not? Students' Explanations for Why They Choose to Enroll (or not) in Chemical Engineering Courses

Abstract:

Most STEM curricula include prerequisite courses which must be completed prior to enrolling in future courses. Prerequisite courses may exist in order to help prepare students for more advanced coursework or to help control course sizes. While courses may be listed as a prerequisite, sometimes courses are listed as prerequisites or corequisites, which gives students the freedom to choose when to take the course in question. Reasons for why students may choose to enroll in a course as a prerequisite or a corequisite have been explored with biology students but to the best of our knowledge this has not been done with engineering students. The goal of this study was to therefore investigate the reasons why engineering students choose to take courses as prerequisites or corequisites and also their perceptions of prerequisite courses in general. Chemical engineering students (n = 162) enrolled in a sophomore level material and energy balances course (MEB) were surveyed about their experience with an introductory thermodynamics course (Thermo), which could be enrolled in as a prerequisite to MEB or as a corequisite with MEB. Students who enrolled in Thermo as a prerequisite (88% of the population) did so primarily because the chemical engineering degree program listed it in the semester before MEB (38% of survey responses) or because it fit well into their schedule (38%), while only a small fraction mentioned the academic value from taking Thermo before MEB (17%). Students who enrolled in Thermo as a corequisite with MEB (12% of the population) did so primarily because of scheduling issues that prevented them from taking it as a prerequisite (59%). Students also perceived that enrolling in Thermo as a prerequisite would correlate with higher grades in MEB compared to students who took Thermo and MEB as corequisites. An analysis of pass/fail/withdraw rates supported this notion as a significantly higher number of students passed MEB when taking Thermo as a prerequisite compared to students taking Thermo and MEB as corequisites. These results suggest that students enroll in courses for different reasons and that we as advisors can provide guidance to help students make decisions about when to take courses as this may impact course performance and retention.

604651 Rapid Transition to Online Teaching: Capturing Chemical Engineering Faculty Strategies, Challenges, and Successes from Spring 2020

Abstract:

In Spring 2020, academia took part in an unintentional experiment as universities cancelled in-person classes across the world. Faculty were forced to move their courses online, and in many cases without access to their campus facilities. In an effort to capture this rapid transition to online teaching we conducted a nationwide survey of chemical engineering faculty at U.S. ABET-accredited institutions. In this talk we report on the timeline of transitions across institutions, types of courses transitioned, and faculty's self-reported descriptions of their course "before" the move to online teaching, and their plans for online teaching. Results comparisons based on characteristics such as institution type (research-intensive, primarily undergraduate, etc.), faculty rank, and demographics to shed further light on how different groups responded to and may have been affected by this transition are considered. Faculty's ultimate execution of their plans, with an emphasis on highlighting challenges and triumphs, are also included. As the academic landscape undoubtedly shifts with recent events, we hope these results will provide both valuable reflective insight into how our community managed these unique circumstances as well as predictive insight into managing future challenges.

609695 Stakeholders Presentations: Presenting to the General Public

Abstract:

Students and professionals will need to present to varied audiences over their careers, but most undergraduate programs focus on presenting to peers and supervisors. We have required our students to make a stakeholders presentation in senior lab. They imagine that their companies want to build a plant that produces or uses a hazardous material. As the company's representative, they must present to city council to convince them that the plant will be safe and beneficial to the community. To prepare the students, we discuss the audience, what they might be expected to know of chemistry and engineering, and how the presentation needs to be different from a technical presentation to supervisor. Expectations and gradesheets will be presented.

599225 Learning to Talk the Talk: Development of an Online Crash-Course in Engineering Communication

Abstract:

To be successful working engineers, students must develop a diverse skillset of both technical and non-technical communication skills. While most chemical engineering students receive training in technical communication through formal reports and presentations, many of the more practical and less formal communication skills are not integrated into the curriculum. In prior work, it was determined that internship students felt that their engineering coursework left them unprepared for communication with non-technical workers, as well as through informal communication such as email, phone calls and online meetings. As a result, an online crash course in practical communication skills was developed to better prepare engineering students for successful communication on their industrial internships. Students receiving credit for their internships were required to complete the online communication course. Course content was developed to teach students about professionalism, teamwork, audience analysis, social and emotional intelligence, and choosing between different communication types (email vs. phone vs. in person communication). Pre- and post-surveys were utilized to better understand the effect of workshop completion on student competencies (skill level, confidence and nervousness) related to communication with different audiences and through different styles of communication. An additional post-internship survey will be used to look at the effect of internship communication on these competencies. Preliminary results indicate that students had the highest level of perceived skill and highest confidence in communicating with engineers. Interestingly, student nervousness was highest with more technical audiences (engineers and managers), and lower with non-technical audiences (non-engineers and the general public). Post-workshop and post-internship data will be collected through the summer and into the fall to determine how competencies change with these student experiences.

593805 Communicating with Both Engineers and Non-Engineers: Incorporating Multi-Disciplinary Communication Strategies in the Unit Operations Laboratory

Abstract:

One of the expected student outcomes for ABET accredited chemical engineering programs is the "ability to communicate effectively with a range of audiences." It can be challenging to design assignments that give students the opportunity to demonstrate this ability, but the Unit Operations Laboratory courses provide a good framework in which to do this. Often, the laboratory courses require some sort of writing, presentations, or other assignments to report results. This paper presents several methods that have been used in a junior-level Unit Operations Laboratory course to practice communication skills. In these assignments, several various audiences had to be considered, from freshman undergraduate students to faculty in chemical engineering. In addition, the assignments are a part of a learning community in which the students have joint enrollment in a technical writing course while they take the lab course, and they must consider the needs of a non-technical audience as well. The assignments will be described and evaluated in light of the desired ABET outcome, and ideas for increasing the opportunities to learn these skills will be discussed.

608351 Reflection Activities As a Pathway for Building Inclusive Team Culture (Work in progress)

Abstract:

Engineering is a collaborative discipline often requiring the use of teams to solve complex problems. The ABET definition of "team" is a group that "consists of more than one person working toward a common goal and should include individuals of diverse backgrounds, skills, or perspectives."[1] Diverse teams in engineering may include people of different disciplines, socioeconomic backgrounds, racial, ethnic, or gender groups, to name a few. Helping teams learn to create an environment that is inclusive becomes very important in engineering education.

As engineering instructors, we encourage teaming and collaboration to solve difficult problems or projects in our courses. ABET places such an emphasis on teaming that it devotes a student learning outcome to teaming: "an ability to function effectively on a team whose members together provide leadership, create a collaborative and inclusive environment, establish goals, plan tasks, and meet objectives." Instructors often expect that students will gain experience about "effective" teaming through the practice of being on a team, without much in the way of instruction on how to be a team member. As a result, students may complete a team experience feeling positive or negative.

When we place students on teams we often neglect to guide them on developing team skills, thinking they'll learn it by doing it. What is lacking is instruction on effective teaming and how to develop a culture of inclusivity and equity. This may lead to underrepresented students within teams experiencing bias, having their tasks shaped by stereotypes, or not having their work or ideas validated[2]. These experiences may impact how they feel about entering a new team experience, and may diminish their self-efficacy and sense of belonging.

This work-in-progress paper explores efforts by the author to use reflective practices to address the development of a "collaborative and inclusive environment." At the beginning of the term, students in teams within the capstone design course sequence are assigned short reflection activities to get them to acknowledge bias[3], map their own strengths[4], and develop team charters. Over time students engage in reflection activities centered on how biases have influenced the roles they are in and how rotating roles can help to develop skills in new areas.

This current work does not specifically address teaching project management skills, including setting goals, planning tasks, meeting objectives, and leadership, but instead focuses on activities that encourage developing a collaborative and inclusive team environment.

To encourage students to engage in and construct meaning from the reflection activities, guidance for each reflection activity was provided as well as a degree of accountability through "graded" journal responses[5]. The activities include reviewing literature that may include topics of diverse teams[6] and/or bias and stereotypes on teams[3], developing personal and team asset maps or matrices[4], creating a team charter, reviewing team charter for stereotyped team roles, rotating roles, and peer evaluations.

In this paper we will share the progress of the pilot testing of these activities and the proposed plan for data collection.

[1] "Criteria for Accrediting Engineering Programs, 2019 – 2020 | ABET."

https://www.abet.org/accreditation/accreditation-criteria/criteria-for-accrediting-engineering-programs-2019-2020/ (accessed Apr. 30, 2020).

[2] B. P. Jackson *et al.*, "Board 124: Interpersonal Interactions that Foster Inclusion: Building Supports for Diversity in Engineering Teams," presented at the 2018 ASEE Annual Conference & Exposition, Jun. 2018, Accessed: Apr. 30, 2020. [Online]. Available: https://peer.asee.org/board-124-interpersonal-interactions-that-foster-inclusion-building-supports-for-diversity-in-engineering-teams.

[3] Lorelle Meadows, Denise Sekaquaptewa, and Marie Paratti, "Interactive Panel: Improving the Experiences of Marginalized Students on Engineering Design Teams. Paper ID # 11803," presented at the 122nd Annual American Society for Engineering Education Conference, Seattle, Washington, 2015.

[4] Elisabeth Stoddard and Geoff Pfeifer, "Working Toward More Equitable Team Dynamics: Mapping Student Assets to Minimize Stereotyping and Task Assignment Bias: American Society for Engineering Education," presented at the CoNECD - The Collaborative Network for Engineering and Computing Diversity Conference, Crystal City, Virginia, Apr. 2018, Accessed: Apr. 30, 2020. [Online]. Available: https://www.asee.org/public/conferences/113/papers/22206/view.

[5] J. A. Turns, G. Scalone, A. Arif, T. Lovins, and C. Atman, "Dimensions in Designing Reflection Activities," in *2017 7th World Engineering Education Forum (WEEF)*, Nov. 2017, pp. 120–125, doi: 10.1109/WEEF.2017.8467101.

[6] Carol Mendin, Megan Lee, and Douglas Bang, "Point of View Affects How Science Is Done," *Scientific American*. https://www.scientificamerican.com/article/point-of-view-affects-how-science-is-done/ (accessed Apr. 30, 2020).

610493 Exploring Social Perspectives Affecting Team Dynamics and Performance in Capstone Design

Abstract:

Engineers in industry work in collaborative teams. Collaborative work and learning in teams involve the study of both group and individual cognitive and social experiences. While the cognitive perspective focuses on the influence of the group work on cognitive outcomes in collaboration, the social perspective looks into the social factors affecting successful performance in group work and teamwork.

The current work will share details on team building activities that explore factors affecting team dynamics and performance in a chemical engineering capstone course using the lens of the social perspective for collaborative work and learning in teams. Research shows that building an effective team includes social skills such as identification of roles, clarification of these roles, agreement upon goals, correct processes and procedures, effective interpersonal relationships, and good leadership. The goal of this work was to introduce the students on team dynamics, what characteristics make a good team member, what team roles are recommended, and some suggestions on how to increase team productivity through better communication. Each team had to produce a team charter that clarified team direction while establishing boundaries. Individual students were surveys twice during the semester, and record of their team progress was recorded to assess the impact of the intervention in team performance. The findings will address class management elements such as group organization, developing team building-vocabulary, specific team-building activities, assessment, and grading.

598765 Capstone Design Team Dynamics in Chemical Engineering at a Hispanic-Serving Institution

Abstract:

Teamwork is an essential part of the capstone design course in engineering. The capstone course is a logical place for gaining extensive experience in a team environment through conduct of the senior design project. Teamwork skills are also crucial for students to develop when they are about to enter their engineering career. Many students at this institution indicate their team experiences prior to the capstone course have been limited and of short duration.

In the capstone design two-course sequence, the students conduct three short team-based projects in the fall semester, and then complete their senior design project in one semester-long project in the spring semester. In the fall semester projects, at least two different methods are used for team member selection, so that the students experience at least one project where they work with team members not of their choice, and perhaps even those students whom they do not know. In the spring semester, students are allowed to form their own teams and usually end up working with others that are friends or study partners. This personal familiarity amongst team members engenders a quick start to their project efforts.

Engineering teamwork experiences are most successful when team members recognize each other's strengths and weaknesses, self-assign tasks in accordance with those characteristics, and step in to help when one person indicates they are having difficulty in completing an effort. The instructor holds weekly meetings with each design group outside regularly-scheduled class time. During these meetings, he questions students on their project efforts and, from this, is able to obtain an indication of who is contributing regularly and who may not be contributing their fair share. Guidance is then given to the students to encourage all members to participate consistently throughout the project term. The individual member contributions to the group are also evident during group oral presentations (four held in the spring semester) and in group member contribution forms submitted twice with deliverables at the middle and end of the semester. The greatest concern is for students who appear to be loners and contribute little or contribute only in an isolated manner. The instructor uses these observations to encourage groups and individual members to participate more fully to ensure that their design project grade does not suffer. However, in some instances, the loner behavior is not corrected and frustration between group members is apparent towards the end of the project. The lone group member scenario is expected to be more prevalent in the current corona virus environment, when face-to-face group meetings are not occurring and most students are participating virtually from home. Numerous groups have recently indicated that the corona virus conditions are a challenge to their completion of group objectives.

606177 The Challenge for Leaders Today

Abstract:

When women are asked the key characteristics of leadership, they tend to cite confidence and determination, but less than half of them feel that they have these characteristics. The challenge for leaders, then, is to instill these characteristics in others through mentoring relationships, especially to young women in academia. Training the next generation of leaders takes time and personal engagement to teach young colleagues how to plan their career, negotiate and find strong role models. Beyond gender considerations, leaders will find that issues surrounding communication and trust will play critical roles in their effectiveness at promoting an inclusive environment.

613110 Manageable Teaching and Service Cheat Codes for Young Faculty

Abstract:

The paths that lead to a faculty position are filled with many valuable challenges and life lessons. This training focusses heavily on producing talented researchers, but to be a successful professor we are also expected to excel in areas of teaching and service. For many graduate and postdoctoral programs, however, there is relatively little focus on developing our skills as educators, administrators, leaders, and mentors. On-the-job training and trial-and-error is typically how new assistant faculty learn to effectively wear these multiple hats.

I now know that, when entering as an assistant professor over a decade ago, my graduate training had not prepared me well for the more human aspects of life in academia. Many hard-won lessons have since helped me find significant and rewarding success, from the departmental to the national level, specifically in those inter-personal facets of faculty life. In this talk I hope to give new faculty the teaching and service "cheat codes" that I wish I had to help prepare for my career in academia.

This talk will focus on employing small steps in teaching and service that I have found to have significant impact. Several key areas in both teaching and service will be addressed:

- *Incorporating evidence-based pedagogy on a budget of time and money.* Effective teaching is a large and complicated topic. But you can easily employ simple and proven teaching methods, particularly interactive and collaborative learning, from day-one in your lecture hall and laboratory.
- *Fostering an inclusive classroom and department.* If a student does not feel welcome in department culture, they are not as able to benefit from our teaching. How do we uphold our AIChE code of professional ethics in areas of diversity equality and inclusion in our classrooms? There are many steps faculty can take to set the stage for a welcoming climate; some of these take little effort but have significant impact.
- *Engaging our communities.* Universities can thrive by incorporating their surrounding communities, and outreach is a powerful tool of faculty service. How can we efficiently use department outreach to attract a diverse student body and strengthen departmental research programs?
- *Being a human professor.* Every faculty is faced with the type of human crises that I would have never anticipated as a new PhD, and our instinct to be "professorial" may be exactly the wrong response. I will end this talk with some tips on how to manage the human distress that is part of being involved with so many people in a very complicated phases of their lives. Being an effective teacher and mentor often means knowing when to *stop* being a professor and stop trying to engineer a solution to our students' problems. In those times, there are simple things we can do to gauge when a student needs help beyond what we are qualified to offer.

612998 The Crazy Learning Curve of Your New Tenure-Track Job

Abstract:

Tenure track faculty juggle an incredible diversity of job skills, most of which need to be learned very quickly and on the go. After perfecting your research skills as a post-doc, you are expected to delegate almost all hands-on research to your team members, instead figuring out how to recruit, build, advise and manage a group. You are now in complete control of the research directions you pursue, which means you are also fully responsible of successes and failures in a new way. You have an empty lab, which needs to be set up from scratch, one stir bar at a time. You are in charge of raising funding to run this group and managing the finances, all while teaching and doing everything else that keeps your department and university running. There are so many new skills you have to cycle through each day, almost all of it brand new. It is quite a learning curve.

I will share my experiences in managing this dizzying process, and discuss the importance of being patient with yourself, keeping an open mind, being flexible, focusing on your students and their growth, and thinking about what you truly value in directing your decision-making. I will also talk about transferable skills I found valuable through my previous experiences as a post-doc, TA/instructor, and a co-founder and senior scientist of a start-up company.

597830 The Dean's Team in the College of Engineering at the University of Akron: Supporting Engineering Recruiting and Outreach

Abstract:

The Dean's Team (DT) at the College of Engineering, The University of Akron (UA), was established in 2013 to support the college's mission in recruiting and community outreach; the program was inspired by a similarly named program at the College of Engineering at Virginia Polytechnic Institute and State University. The DT is comprised of about 25 or so mostly undergraduate students within the College of Engineering and distributed across the eight programs offered within the college. The DT helps the college with recruiting and marketing in several ways. First, the DT represents each program at UA structured visit days. For the Department of Chemical, Biomolecular and Corrosion Engineering, the DT members share their experiential learning experiences and job functions as a co-op student during departmental presentations. Prospective students and parents appreciate their first-hand stories and their views of engineering studies and engineering careers. Secondly, the DT hosts individual visits in-person and virtually. This one-on-one interaction helps prospective students become familiar with the details of student life and program requirements. The DT also reach out to the students and families after visits through notecards or emails. Some of the DT members also offer hands-on workshops to high school students, along with panel discussion about college life and engineering design projects. The critical factors to the success of the DT program include identifying the best students for the team and appointing someone to train and manage the team. For 2019 Fall recruiting, with the support of the DT, the yield rate at the College of Engineering is about 11% higher than the university's average. Not only did the DT help with recruiting, but also eased students' transition from high school to college.

In this presentation, we will provide a history of the Dean's Team, the costs (both personnel and financial) of running a program, the impacts of the program and how the Dean's Team members are chosen and utilized in the college, and specifically within the Department of Chemical, Biomolecular and Corrosion Engineering at the University of Akron.

600892 Putting Students First: Building Quality Academic Advising from the Ground up

Abstract:

Academic advising is an incredibly important part of an undergraduate student's experience and contributes strongly to their success. There are numerous models for schools and departments to consider, including faculty-based advising, professional staff-based advising, and hybrid models that combine elements of both. As enrollments in many engineering programs grow and engineering becomes more multidisciplinary, programs need to consider how to best meet the advising needs of their students. Many see academic advising as focused on coursework or other purely academic topics, but advisors also counsel students on their professional development, career choices, and in many cases, personal lives as well. As such, developing a skill set that treats students more holistically and also allows the advisor to connect students to other resources on campus is imperative. In this presentation, we discuss the best practices we have observed and built as the director and associate director of advising at a large, public, R1 engineering program, where we have focused on student success, student self-actualization and agency, and building collaborations with faculty in the school and with resources across the university.

609955 Pre-Professional Student Groups and Your Department: A Partnership

Abstract:

Every department and college have several of them: Society of Women Engineers (SWE), Engineers without Borders (EWB), National Society of Black Engineers (NSBE), Women in Science and Engineering (WISE), Tau Beta Pi ($\tau\beta\pi$), and American Institute of Chemical Engineers (AIChE). These are all student pre-professional groups, and they serve as student support programs for a variety of different purposes: mentoring, scholarship, service, and learning about the student's future profession. Each of these student groups have their own separate missions, activities, leaders, and needs. Each group has a regional or national conference for students or a need to travel to serve others in need. Inevitably, each group will be visiting the department head or Dean to ask for financial support for their travel and other activities. The department head and Dean are happy to help where they can, but they might also ask what do these student groups do for our department and college? Can we find a way to build a better partnership between the student groups and the department/college to better meet all of our needs? Can we help them? Can they help us? Can we help them help themselves?

So what is the relationship between the department/college and these student groups? The student groups need our money and advice, and we need their labor. Thus, building a partnership benefits both sides. Is the department/college looking for content for their website and newsletters? Are prospective students looking at the website to see student activities that will entice them to apply and enroll? Do the student groups need good advising and financial resources to help them travel? Are the alumni willing to provide donations to help the students and student groups? Can we publicize their activities to help our recruiting? These mutual needs and capabilities provide the basis for a building a strong partnership.

In this presentation, the partnership between these student groups and the department/college to improve these support programs will be discussed. Partnerships between the units and the students can lead to:

- more vigorous student groups,
- additional mentoring and recruiting avenues,
- better and more frequent publicity for the units,
- more opportunities for students to get involved,
- more student leadership opportunities,
- mentoring to increase student abstract submissions to conferences (and the subsequent presentations),
- more connections with the university fundraising arm for more marketing and advancement opportunities for the student groups and units,
- better student/alumni relations, and
- more opportunities for faculty to know the students and their skills which leads to more successful recommendation letters for the students.

Start looking at student groups as partners to support the unit's mission. This partnership actively promotes the students and their groups while helping the department or college.

599992 Impact of a Computer Programming Support Center on (Chemical) Engineering Students' Success and Satisfaction

Abstract:

In the chemical/biochemical engineering curriculum at the study institution, 11 required undergraduate courses taught by the Department of Chemical Engineering include a substantial programming component, using Matlab, Python, and/or process simulation software (Aspen Plus™ or SuperPro Designer®). In response to demand for increased out-of-class programming support specific to these courses, an expansion of existing support specific to community college transfer students was implemented in January 2020. While the focus of this program was on (bio)chemical engineering students, this support was made available to all engineering students at the study institution. The ongoing goals of this project are as follows:

1. To retain at risk undergraduates in difficult, programming intensive courses, keeping students on track for a 4-year graduation (2-years for transfer students).
2. To increase overall student success and satisfaction in programming intensive courses.

In response to an earlier study done on the barriers to success for transfer students in engineering at the study institution [1], a programming crash course in Matlab, piloted in Fall 2017, was developed to address the lack of programming preparation cited by students, and in particular those taking courses in the Department of Chemical Engineering. This crash course has been offered to incoming transfer students in the Fall of their transfer year as an optional 1-credit course for the past three years. Survey results and anecdotal feedback have shown that transfer students find the course to be valuable (course's educational value rated 4.5/5). Additionally, it was found that a number of non-transfer students were interested in the course due to their perceived lack of programming preparation for their junior-level courses. However, despite the interest and positive feedback, students have found it difficult to complete the crash course requirements while also balancing their heavy load of required courses. What had emerged from the three iterations of the programming crash course was that both 4-year and transfer students were eager for on-demand programming support, but hesitant to commit to or persist in a formal course. Additionally, this need for programming support goes beyond the junior-year Fall quarter and transcends the entire (bio)chemical engineering upper-division curriculum. In addition to the survey and anecdotal feedback related to the crash course, students have been demanding additional support on their senior exit surveys for not only their Matlab/Python intensive courses but also for their courses that heavily utilize process simulation software (Aspen Plus™ or SuperPro Designer®). In Spring 2017, the junior class at the study institution was also surveyed and their lack of training and support for their programming intensive coursework was also heavily cited.

In order to address this need for additional programming support for students, the Department of Chemical Engineering at the study institution established a team of graduate teaching assistants and undergraduate tutors under the direction of a faculty director who were collectively responsible for providing both one-on-one tutoring sessions and small group programming support sessions. In order to assess the goals of this project, a number of direct and indirect assessments were carried out. These included:

- Direct Assessments (indicates when assessed)
  - Grades in target courses of students who use services vs those that did not (quarterly)
  - Overall average GPA of class in target courses vs previous years (quarterly)
  - Number of students and demographics of students who used services (quarterly)
- Indirect Assessments (indicates when assessed)
  - Survey of students who use services, to include satisfaction with services, perception of gains in programming skill level, and sense of satisfaction with target courses (quarterly)
  - Comments on senior exit surveys related to programming skills and target courses as compared to previous years (annually)

In addition to answering the demand for out-of-class programming support, multiple benefits to both our students associated with this project were anticipated. The first expected benefit was that students would persist in difficult, programming intensive courses instead of dropping or delaying courses thereby delaying their graduation. Secondly, students who took advantage of this support program were expected to see improvement in their performance, particularly in the target (programming intensive) courses. Further, by becoming more adept at using these programming tools, students were expected to have a greater sense of satisfaction with respect to their programming abilities, their performance in these courses, and in their overall college experience.

References:

1. Gentry SP, Bronner CE, Choi JH, and White JR. (2018) "Successes and Difficulties Experienced by Engineering Transfer Students at a Large Public University." *Proceedings of the 2018 ASEE Annual Conference and Exposition.*

595875 Integration of Professional Development Practices in Doctoral and Postdoctoral Training – Approaches for Supporting Career Navigation in Chemical Engineering

Abstract:

Skills such as project management, collaborative problem solving, negotiation, conflict resolution, technical communications, personnel management, finance, and time management are critical for STEM careers. Such skillsets are predominantly acquired by graduate students through primary interactions with their thesis mentors or are self-sought externally, yielding variability in student preparedness for the workforce post-graduation. We hypothesize that standardization of professional development for STEM graduate students will enable graduate students to hone these skills prior to entering the workforce, and could decrease the achievement gap of underrepresented students. In this talk, we will discuss standardization of professional development within the context of a Chemical Engineering Ph.D. thesis via strategies to benchmark student and postdoc progress in individual research groups. Methods to establish a positive mentoring relationship, assess progress through Individual Development Plans (IDPs), and tailor the Ph.D. or postdoctoral training period for various target careers are discussed. Furthermore, we disseminate resources for standardizing the assessment of undergraduate research and how to train graduate students and postdocs in the skills of project management and mentorship. Resources presented in this talk will be made broadly available for group leaders and primary investigators wishing to implement professional development practices in their research groups.

611686 Advocacy for the Tenure-Track Teaching Professor

Abstract:

"Lecturers do not perform academic research and hence are not real professors." This is a common sentiment from engineering schools around the world. But these faculty can provide many services beyond that of a temporary lecturer. In this talk, an argument is made in favor of providing tenure-track positions to lecturers. Beyond teaching any and all core courses, these teaching faculty can perform pedagogical research, create highly specialized elective courses, provide academic advising for students, and develop higher quality unit operations laboratories. Teaching faculty can even provide new sources of income for departments by developing new graduate programs or maintaining alumni networks. Evidence of success will be provided by highlighting the University of California's Lecturer with Security of Employment (LSOE) position.

599967 Creating an Inclusive Chemical Engineering Community for Womxn through Allyship and Advocacy

Abstract:

Women in Chemical Engineering (WChE, "Wick-ee") was founded in January of 2016 with the primary purpose of educating, empowering, and advocating for women in chemical engineering, and their supporters, by providing a safe and open space for dialogue, mentorship, collaboration, camaraderie, and skill building through professional development, community networking, and outreach to primary, secondary, and post-secondary institutions. The founding chapter began at the University of Washington (UW) and consists of undergraduate, graduate, faculty, staff, and professional women, and all-gender allies, in chemical engineering. A second chapter was recently founded in 2018 by Beverly Miller at the University of Virginia (UVA), and a third chapter, started by Belinda Garana, came online in 2019 at the University of Southern California (USC).

In this talk, we will first review the data motivating the foundation for WChE. We will then discuss how UW, UVA, and USC have built WChE chapters to foster an inclusive chemical engineering community for womxn. We will highlight many of the events WChE chapters have created, organized, and hosted, ranging from "How to be An Ally" panels to "Introduce a Girl" outreach series in partnership with the Girl Scouts of America, to Annual Industry Events. Each event serves one of the three WChE pillars: professional development, community development, and outreach. WChE activities and events are organized to meet the organization's guiding principles: (1) to promote women in chemical engineering while not furthering self-stereotyping or stereotype threat; (2) to increase and elevate the presence of women voices at the table as critical contributors to drive forward thinking, high-impact change; (3) to further inclusion by creating a space for all identifications of genders to engage and interact through allyship and advocacy. Additionally, in partnership with the Women in Chemical Engineering (WIC) AIChE community, we aim to establish WChE Chapters at more universities, and will provide a brief overview of our "WChE start-up package."

610561 Student Driven Design Problems

Abstract:

For design courses it may sometimes be difficult to author challenging and interesting problems for the students that cover all the needed concepts for these culminating experiences. While there are a wealth of AIChE competition problems available, often solutions to these problems are discoverable, or may not fit with contemporary issues. What has been evaluated at Kettering University is a student-driven process in devising the context of a design problem. This process starts with students reviewing current events and safety topics, Grand Challenges, global issues, and items of interest. Each student is provided a few minutes to give a presentation on their idea for a design project and these are down-selected to handful of promising problems. Next student teams are formed to further evaluate the promising problems and determine the depth of the problem. The teams then present their findings in twenty minute presentation. Following this the professor will ultimately select a topic from which a culminating design problem may be synthesized. The end result is the class is not only invested in the process of determining the theme and some of the parameters of the design problem, but ultimately preparing a solution to a problem of they are motivated to solve. Problem encompass not only lifelong learning exercises, but study into process safety. Examples of design problems include synthesis of return rocket fuel for use on Mars, production of transportation ethanol, production of ethanol for consumption using renewable energy and capture of carbon dioxide, and desalination of seawater for drinking water using renewable energy for the city of Irvine, California. Examples of the problem generation process and design problems will be presented.

604889 Hands-on Biochemical Production and Process Design

Abstract:

Biochemical production offers a renewable method to generate a variety of small molecules for use as fuels, polymer precursors, scents, flavorings, colors, and pharmaceuticals. However, few processes have made it to full-scale production due to competition with conventional chemical synthesis. We have tasked students to evaluate the economics of scaling up biological production of their choice of three molecules: violacein (a purple dye with antimicrobial properties), psilocybin (a hallucinogen in clinical trials for treatment of depression), or caffeic acid (a nutraceutical with antioxidant properties). This capstone project combines on-paper process design with laboratory-based parameter measurement to offer a hands-on project where students build skills in microbial fermentation, bench-scale purification, and chemical analysis.

Students begin their project by conducting an economic analysis to identify a target market, competing technologies, selling price, and yearly production amount for their molecule of interest. Based on this analysis and a literature review, students create a preliminary process design that meets their specifications (i.e. productivity, purity) as well as identifies key parameters that will influence their scaling. Next, teams enter the laboratory to learn about recombinant bioproduction and to measure the parameters needed for their proposed design. A key aspect of this project is that students grow robust and well-characterized bacterial strains and use known chemical analysis methods. This enhances the feasibility and success of the experiments, especially for students with limited laboratory experience, by largely eliminating unexpected and time-consuming issues associated with cell growth and product analysis. Further, downstream processing through bench-scale drying and extraction experiments are completed to provide yield, stability, and purity process parameters. Finally, the empirical data is combined the original design to create a full process design with scaled unit operations to meet their target market.

Our design project has been offered for the past two years with 18 students participating in groups of 4-5 students in a two-semester capstone design sequence. Through pre- and post-test data we have observed enhanced student learning in microbial fermentation, chemical separations, and molecule analysis techniques. Students are engaged due to it being a hands-on project that combines many topics taught in previous engineering courses to design a process for molecules that benefit society. We have found that this design is highly complementary towards traditional chemical process design and biochemical/bioprocess engineering courses as well as builds on topics covered in separations, kinetics, and material and energy balances. To share this course, select bacterial strains and culture methods will be made available. Student-collected data will also be shared for each biomolecule for fully on-paper design projects.

611268 Process Design and Critical Thinking

Abstract:

For the last six years, students in the capstone process design course are required to design a process, which converts raw materials into a desired product. This process design is completed in teams of 4-5 students and each team designs a different process. The design assignment is split into different sections, which include; process synthesis and balances, equipment design, process optimization & impact of design on global, economic, environmental and society, and process economics. In order to provide our students the opportunity to effectively communicate their work, critically analyze and critique open-ended designs, the projects are rotated between teams for each section of the design. During the rotation of projects, the teams have the opportunity to evaluate one another's work, integrate design reviews into their final analysis and learn by critically analyzing other projects.

This is the sixth year of this approach to teaching the capstone process design class. This presentation will include specific details/examples of course/team logistics, team and individual assessment and direct assessment of learning objectives. In addition, the technical evolution of the design will be quantified by examining the design changes in response to the critiques provided by the other teams.

597930 Capstone Design Twitter Poster Session As a Way to Engage in Discussions and Feedback Online

Abstract:

As part of a senior Capstone Design 2 course, a poster presentation was previously done to allow teams to discuss and get feedback for their detailed process designs before their final designs are submitted. This poster session was designed for an online format using the social media platform, Twitter. The 23 design teams posted their posters using the hashtag #CHME4703Poster and had 24-hours to comment and discuss their project designs. The use of Twitter as a social media platform enabled engagement, discussions and feedback on their designs by the Chemical Engineering community from their peers, faculty, and alumni at our University as well as the larger scientific and engineering community, notably through AIChE and ASEE. The event was a success with over 275 tweets, 382 likes, and 81 retweets. Students commented that they liked the online Twitter format as a way to engage and get the feedback from the Chemical Engineering community similar to they would in a face-to-face poster session.

607670 Design in and out of Capstone

Abstract:

This talk will briefly present and share materials for three chemical engineering design projects spanning the curriculum. First, I will share a first-year product design experience and give an overview of its many variations (sneakers, salad dressing, swimming pools, among others). These experiences are designed to occur over three weeks of combined classroom and laboratory meetings, but may be altered for a one credit hour semester-long intro course or summer pre-college class.

I will also share a chemical process design "game" that can serve as a fun motivation in reaction thermodynamics. This module is useful as a preview of upcoming capstone design, as it asks students to balance costs for raw materials and plant conditions with potential sales from chemical products in an equilibrium-limited reaction, the creation of syngas by steam reformation of methane. Students compete in class (or virtually) to have the most profitable plant in this design optimization exercise.

Finally, Bucknell's approach to the classic capstone chemical process design and process modeling course will be presented. In this course, student teams are formed before the start of class, and given a local feedstock, most recently from the Marcellus shale. Teams propose a product, a reaction pathway, and use Hysys to simulate the plant including separations and energy integration. Teams conclude with a feasibility report including economics.

611554 Steal This Curriculum-Wide Design Project: The HDA Process

Abstract:

Imagine a design project that spans not just a single term, but the full curriculum. Students are introduced to a full industrial process in their first course, assuming ideal systems and identifying points of interest for future design optimization. In every subsequent course, the students analyze the same process - modifying, adapting, and incorporating that course's subject material. By the time the students are in their senior capstone design course, they've already been exposed to the full design protocol and are ready to apply that knowledge to more complex or modern processes. In this talk, this vertically integrated curriculum design project is outlined in context to the HDA process for benzene production.

605661 Choose Your Own Kinetics Adventure: Student-Designed Case Studies for Chemical Reactor Design Projects

Abstract:

This work presents an innovative approach for a chemical reaction engineering course project. The project tasks students to conduct the preliminary design of a reactor or series of reactors to produce a chemical product of their choice. The constraints on this open-ended "choose your own project" are that the chosen process must involve a catalyst, multiple reactions, and heat transfer. The students must use published kinetics and physical property data from the literature, heat and mass balances, and software to complete their design. Students work in groups and write a report summarizing their findings. The activity connects course concepts to real world applications and requires students to design their own case studies through exploring the research and patent literature. These aspects engage students in topics they are interested in while simultaneously relieving the burden off of faculty for constructing new projects each course offering. This project has been employed successfully in six offerings of Chemical Reaction Engineering at Oklahoma State University, including during Spring 2020. Only minor modifications were needed to adapt this project for use in the online course format. This type of project in a junior level course prepares students for the senior capstone design experience the following year.

609814 Active Learning Exercises for Teaching Reactor Design

Abstract:

Dr. Ashlee N. Ford Versypt has taught Chemical Reaction Engineering (CRE) for chemical engineering juniors at Oklahoma State University for six years. She also co-led the AIChE Education Division's Virtual Community of Practice on CRE & Process Control during Spring 2020.

Formulating and solving complex reactor design problems is a primary learning outcome of the CRE course. Active, problem-based learning approaches are used to build a strong basic understanding and then progress incrementally to more demanding concepts and problems. She employs a modified flipped classroom approach. Students are required to read chapters and supplemental readings before topics are covered in class. Most of the in-class time is spent discussing how to work problems, doing examples, and giving student opportunities to work problems while they can quickly get her feedback and interact with peers on learning the material. This framework is referred to as "I do, we do, you do." First, Dr. Ford Versypt presents a case study and works a realistic example. After showing how she approaches solving a particular type of problem, the students form teams of two or three and work together in class to solve a similar problem that is slightly more challenging or is intended to unveil a commonly misunderstood point. While students work, she circulates to answer questions before the group comes back together to debrief. Often students are encouraged to explain their methodologies or pitfalls instead of just having the instructor tell the answers. Finally, students use these same problem-solving skills on their out of class homework assignments. The course is structured so that the problems that the students see on assessments lead toward achieving the course learning objectives.

In this presentation, examples of active learning exercises and memorable analogies will be shared for other chemical engineering educators to adapt for their own courses. The course opens with analyzing the illustration of witches brewing a potion that was the cover of Octave Levenspiel's The Chemical Reactor Omnibook as a way to brainstorm the course topics to be covered during the semester. One exercise uses grilled cheese sandwiches to demonstrate stoichiometric tables. Another analogy relates the mass transfer and reaction processes involved in heterogeneous catalysis to the plot of the '90s blockbuster movie Armageddon. These exercises culminate in a semester group project where students apply course concepts to design the reactor(s) for synthesizing a chemical product of their choice.

AUTHOR INDEX